BATTLE HYMN OF THE DISREGARDED

FROM FOXHOLES TO COURTROOMS - THE LEWIS CAIN STORY

BY

Dr. Kara Lock-Harris

This memoir is based on the life and experiences of Lewis Cain. To protect the privacy of individuals and to allow for narrative flow, certain names, locations, dates, and identifying details have been changed or combined. Some dialogue, scenes, and events have been reconstructed or dramatized from memory, interviews, public records, and creative interpretation.

The author and publisher make no representation that this work constitutes an exact or complete account of any real-life events. It is intended for literary and educational purposes only. Any resemblance to actual persons, living or deceased, or actual events is coincidental or used fictitiously.

All opinions and interpretations expressed herein belong solely to the author and contributors.

The author and publisher disclaim any and all liability arising directly or indirectly from the use or interpretation of the material presented. Legal proceedings, correspondence, and professional interactions described in this work are presented as recalled and interpreted by Lewis Cain from his personal experience and documentation. They do not constitute findings of fact, legal conclusions, or allegations against any identified individual.

ISBN: 979-8-9930156-0-6 (Paperback)

ISBN: 979-8-9930156-3-7 (eBook / Digital)

First Edition, 2026

Library of Congress Control Number: 2026911725

Printed in the United States

Dedication

This dedication reflects the heartfelt wishes of Mr. Lewis Cain, who asked that this book honor the family whose love shaped his life and whose sacrifices carried him through every battle he has ever faced.

It is dedicated first to his mother, Mary Lindsay Cain, whose prayers rose early in the morning and wrapped themselves around her thirteen children with the strength of a shield. She built their home with her own hands and built their character with her faith, teaching them to love, to serve, and to stand for what is right.

It is also dedicated to his siblings, both those who came before him and those who came after. Each one is part of the foundation he stands on. Those born before him carried responsibilities greater than their years and made sacrifices that helped the family survive. Those born after him continued the unity their mother instilled, supporting one another with the same spirit of love she taught them.

With deep love and honor, he acknowledges his brothers, Lamar Cain, Leroy Cain, Larry Cain, Walter Lee Cain, Bennie Cain, George Davis, and Clarence Davis, whose lives and presence remain an enduring part of his journey.

In honoring them, we honor the legacy that shaped Lewis into the man he is. His story stands on their shoulders, and this dedication belongs to them, just as he wished.

Contents

Introduction

Some stories demand to be told, not just because they are extraordinary, but because they echo the quiet struggles and loud injustices that so many people face, yet rarely see honored in print.

This is the story of Lewis Cain. But make no mistake, it is also the story of every person who has ever been told they don't belong, every person who has been denied a seat at the table, every person who has been told to wait their turn, lower their voice, or settle for less. It is the story of survival against the odds, of standing back up when the system tries to knock you down, of holding onto dignity when the world seems determined to strip it away.

Lewis Cain's life is not just a veteran's story, though his service in Vietnam forged in him a toughness few of us can imagine. It is not just the story of a man who filed legal complaints. However, his courtroom battles and struggles with bureaucracy reveal a deep and painful truth about justice in America. This is the story of resilience. A story of what it means to live as a Black man in a country that promises equality but delivers injustice. Of a man who has been poisoned by war, denied by his government, betrayed by his employers, robbed by the systems that should have protected him, and who still refuses to be silent.

Lewis Cain's life is not just a veteran's story, though his service in Vietnam forged in him a toughness few of us can imagine. It is not just the story of a man who filed legal complaints. However, his courtroom battles and struggles with bureaucracy reveal a deep and painful truth about justice in America. This is the story of resilience. A story of what it means to live as a Black man in a country that promises equality but delivers injustice. Of a man who has been poisoned by war, denied by his government, betrayed by his employers, robbed by the systems that should have protected him, and who still refuses to be silent.

This book is a tribute to his fight to stand. It is a tribute to every person who has ever felt like giving up, and a reminder that sometimes the greatest victories are not measured in court wins or settlements, but in the refusal to give up one's voice, one's story, one's worth.

Lewis Cain's journey matters. Not just to him, but to every reader who picks up these pages and recognizes their own reflection in the struggle. This is a story about pain, yes, but also about power. About faith. About justice. About the fight to be seen, and the courage it takes to keep standing.

Chapter 1: Coffee with Jesus

"I did not know Lewis yet, but we were already sitting at the same table. Two cups. One Savior. And a story heaven had been brewing long before we spoke."—Dr. Kara

The morning it happened began like any other. My Bible sat open beside my cup of coffee, the steam curling upward the way prayers rise when words fail. I had been talking to God quietly, asking Him to give me strength for the battles I was facing behind the scenes. Personal storms were gathering, and although I stood in faith, my heart was tired. Still, I breathed in the quiet of the morning, grateful for the few moments of peace before the demands of the day arrived.

I opened my laptop almost absentmindedly to check messages. Nothing felt out of the ordinary. Nothing hinted that my life was about to shift. Then, a single Facebook notification appeared.

I almost ignored it.
Almost.

But something in my spirit nudged me to open it.

The message was simple.

"Hello, the service I need help with is a ghostwriter. I have glaucoma, and my eyesight is not as good as I need it to be. Can your business help with this? Thank you."

I read it twice. Something about the directness caught my attention. Before I could respond, more messages came in. Long messages. Emotional ones. Messages from a man who clearly had been carrying too much for too long.

"Good morning to my business solutions partner. I think I have an interesting, unusual story. The reason that I am up so early this morning is that I do not sleep well. I am an eighty-year-old Black man. I was the first ground troops in the Republic of South Vietnam. We were sent there by President Lyndon Johnson in March of 1965."

I sat up straighter in my chair. His words carried significance. They did not read like a business inquiry. They read like a door opening into a life that had survived storms I could not imagine.

He continued.

"I want you to write this story about the things that this old school Black Marine Corps veteran has to deal with every day of my life. I am not complaining. Not at all. I am a fighter, and I am grateful to God for His mercy and grace. I was hoping you could write a book and have it made into a movie. You get all the credit and go on to become well-known worldwide. I just want to help you on your journey."

I stopped reading. My breath trembled for a moment. I whispered, "Lord, what is this?"

This stranger, this man I had never met, was speaking purpose into me with a confidence that startled me. I scrolled further.

"I think I have been waiting for you to get here and to do this. I just had coffee with Jesus. We talk every morning, and I know your spirit is good and ready to go."

My hand covered my mouth.

Coffee with Jesus.

The same phrase that shaped my personal devotion time. The same phrase that anchored my blog, "Letters From Christ Blogging Over Coffee." The same ritual that steadied me through the private storms I never told anyone about. I felt God's presence settle around me like a warm blanket. This was no coincidence. This was alignment.

He wrote more.

"The only thing I want you to do for me is that your work will be dedicated to my mother, my sister, and my brother who came before me. They are the reason I am here. God used them to help all my younger siblings and me. My mother had 13 children: 9 boys and four girls. We have always helped each other and anyone we could help. That is how our mother taught us. To love. Because it did not hurt."

I exhaled slowly. His story was not just a story. It was an inheritance. And he was offering me the chance to help deliver it to the world.

He mentioned a court hearing he was headed to that same morning.
"The Harlow Falls Village Hall," he said. "The city attorney's office is taking me to court about the well that's on my property. I have to deal with that too."

His words lingered with me. In a single breath, he revealed yet another battle he was fighting, one more burden added to a life already carrying far too many. It humbled me. One man. One message. And somehow an entire lifetime of struggle woven between the lines.

I responded to him, and later that day, I dialed his number. He answered on the second ring.

"Hello."

"Hi, may I speak with Mr. Lewis Cain?"

"Speaking."

Even in those first two syllables, I could hear it. His voice carried history. There was age in it, and pain, but also strength and a quiet gentleness that pressed through the fatigue.

"Good morning, sir. This is Kara."

"How are you doing?" he asked warmly.

"I'm doing well. How are you this morning?"

"I am okay. I cannot complain about anything."

But he could have. Anyone in his position would have had every reason to. Yet he chose gratitude, not because life had been kind to him, but because his spirit refused to bow to bitterness.

But something about Lewis was different. He carried a quiet dignity that did not allow self-pity.

"I want to start by saying thank you for this opportunity," I told him. "I am honored, and I do not take it lightly."

"I thank you," he said. "I am just trying to leave a message behind. Something that says I was here."

His words filled the space between us. Gentle. Heavy. Sacred.

"I understand," I whispered.

He began sharing pieces of his story, not in bullet points or rehearsed lines, but in the gentle, weary flow of someone who had carried far too much alone. He talked about the discrimination he endured, the injustice that trailed him from courtroom to courtroom, the stacks of documents he had saved over decades because truth had become his only shield. His pain surfaced not through anger, but through the tired steadiness of a man who had been denied the rights promised to every American.

"For thirty-five years," he said quietly, "they have denied me. But I keep everything. I have it all documented. I just want justice."

His voice did not shake. It did not crack. It simply rested in that sacred space where truth becomes heavier than grief.

We drifted into conversation about his upcoming Channel 4 interview and the contamination at Camp Lejeune that poisoned his body. He mentioned the doctors, the health battles that never seemed to end, and his eyesight fading because of glaucoma. He spoke about his brothers and sisters, about sacrifice and survival, about how faith had kept him from collapsing under the weight of it all.

Then he said something that made the room feel strangely still.

"I want you to get all the credit. I just want to help you on your journey."

I closed my eyes for a moment. His words moved through me like a warm current, unexpected and sincere. I felt something inside me soften, something in me that had been tightening under the pressures of my own life. I whispered inwardly, "Lord, I think I need this as much as he does."

He went on to tell me that he believed God had brought us together and that he had been waiting for someone to help him tell his story. As he spoke, something in my spirit shifted. It was subtle at first, like a quiet click inside a lock, but I knew it was God. Something larger than both of us was unfolding.

We scheduled our first in-person meeting for Saturday morning.

"Ten o'clock is fine," he said with certainty. "I will be ready."

And I believed him. Not because of his tone, but because something in the air told me this meeting was never about convenience. It was about destiny unfolding at just the right time.

Before hanging up, I asked, “Mr. Cain, would it be all right if I close our call with a short prayer?”

“That will be fine,” he said softly.

I lifted my head upward towards Heaven.

“Lord, I thank You for Mr. Cain. Thank You for sustaining him through every battle he has faced and every injustice he has endured. Father, cover his life with Your peace. Surround him with Your angels. Protect him from dangers seen and unseen. Strengthen his body, quiet his mind, and bring justice where it has been denied for far too long. Let this book become the legacy You intended and let it reach every heart that needs truth, healing, and hope. Guide us, protect us, and let everything we do honor You. In Jesus’ name, amen.”

He exhaled softly.
“Amen,” he whispered.

There was a tenderness in his voice when he added, “You know that song, Make a Way, by Tramaine Hawkins? That is my favorite song.”

I smiled. “I believe God is making a way right now.”

After we hung up, I sat still for a long moment, holding the cool mug of coffee in my hands. The morning felt different now. Sacred. Appointed. Heavy with destiny.

I had been carrying my own battles quietly, trying to stand through storms I could not speak about yet. But hearing Lewis talk about his life, hearing his resilience, his faith, his unwavering determination to survive, awakened something inside me. His voice reached places in me that had begun to feel weary, reminding me that God was still writing my story too.

Saturday came quickly.

I parked outside his home in Harlow Falls and felt the significance of the moment settle onto my shoulders. I did not know what I was about to walk into. I did not know how deeply his story would reach into my own. But I felt God in the air as I approached his door.

Lewis opened the door slowly, and for a moment we simply stood there, taking each other in. He was taller than I expected, standing with a quiet, disciplined posture that spoke of the Marine he once was and the battles his body still carried. His frame was thinner, his movements careful, but his eyes... his eyes held something fierce. Pain lived there; yes, years of it layered in the way he blinked and the way his breath settled. But behind it burned a fire that refused to be put out. A Marine's fire. The kind that even war could not kill.

"Come in," he said, stepping aside.

I walked inside, and the atmosphere shifted immediately. His home felt like stepping into a living archive. Not a museum of objects, but a museum of meaning. The walls were filled with stories long before he spoke a word.

Photographs lined every angle. Some framed in polished wood, others in simple glass. In one picture, a young Lewis stood beside his mother, her expression strong and steady, the kind of woman whose spirit could move mountains. He saw me looking and said softly, "That is my mother. Everything I am, came from her."

Another photo showed his brother in uniform, standing tall and proud. "He served in Vietnam too," Lewis told me, his voice warming with both pride and grief. "We all had battles to fight."

He pointed toward a cluster of medals mounted neatly beside certificates from the Marine Corps. Bronze. Silver. Ribbons of red, white, and blue. Each one represents sacrifice. Pain. Honor. Survival. I stepped closer, running my eyes over engraved names and dates. "These are beautiful," I whispered.

"They came with a cost," he replied.

His tone carried no bitterness, only truth.

A folded American flag rested inside a triangular frame. He touched it gently with two fingers, the way someone touches a memory that still aches.

As I continued looking around, something in the living room drew my attention. An exercise bike stood in the center of the space, not tucked away, not hidden, but placed boldly in the middle as if it belonged there more than the furniture did.

He noticed my eyes drift toward it. “That right there is how I keep my legs moving,” he said. “They hurt me most days. The pain gets bad, but I gotta keep going. If I stop, everything gets worse.”

I nodded, feeling the load of that sentence.
He was not working out for fitness.
He was working out for survival.

Every detail in that home whispered resilience.
The medals.
The certificates.
The folded flag.
The photographs of the people who made him who he was.
And the exercise bike that stood as a testament to the fact that he had no intention of giving up.

He motioned for me to follow him into the kitchen. The table waiting for us was simple and clean, almost sacred in its emptiness. Two chairs. Two stories. Two paths God had somehow woven into one.

We sat across from each other. I set my phone down.

"Is it all right if I record our conversation?" I asked gently.

"Yes," he said. "You record everything."

I pressed record, expecting our conversation to follow the outline I had carefully prepared. My notebook sat open beside me, filled with pages of typed questions arranged in neat categories. I had done my homework. I had a plan. I thought I knew how the story would unfold.

I started with the question I always ask, the one that opens the soul in a way no icebreaker ever can.

"What can you tell me about your childhood that shaped who you are today?"

Lewis didn't answer right away.
He sat back. He breathed.
And then something in him shifted, like a door unlocking from the inside.

When he finally spoke, his voice didn't stay in the kitchen. It reached back through decades, pulling the room with it. Suddenly, I wasn't looking at an eighty-year-old man across a table. I was hearing the echo of his younger self, running barefoot through Mississippi soil, surviving things most grown men would crumble under.

His voice carried the substance of red dirt roads, Vietnam humidity, injustice layered on injustice, and the resilience of a man who had been fighting long before he ever picked up a rifle. With every sentence he spoke, the air around us deepened.

My breath slowed.
My posture leaned forward.
And my heart began to pound in quiet disbelief.

The questions I prepared no longer mattered.
His story had its own heartbeat.
Its own rhythm.
Its own urgency.

Words poured from him with a force that felt ancient, as though they had been waiting all this time for the right witness. He didn't stumble. He didn't hesitate. His memories came alive, rising with so much clarity it felt like watching a film through his eyes.

By his fifth sentence, I knew the truth.

I had absolutely no idea what I was about to hear.
And nothing in my spirit, my training, or my personal storms could have prepared me for what was coming.

Lewis took a slow breath, his gaze drifting somewhere past the kitchen window, back into a time he had never fully escaped.

“It started in Mississippi,” he said.

The moment the words left his lips, something in the room shifted. It was subtle at first, like a change in temperature, a quiet tightening in the air. Mississippi wasn’t just a place he was naming. It was the ground his entire life had grown from. It was the soil of his earliest battles, the origin of wounds that never fully healed.

He didn’t say the word with anger.
He didn’t whisper it with fear.
He spoke it the way a man speaks about a truth he has carried for far too long.

Mississippi.
The place where danger first learned to call him by name.
Where pain and purpose tangled together before he ever set foot in boot camp.
Where a boy had to become a survivor long before the world called him a Marine.

Those four words carried heat and memory and something like fire. As he spoke them, I felt myself being pulled into the red clay of his childhood, into the world that shaped him long before war ever did.

The kitchen around us faded a little. Not literally. But in a way, memory stepped forward and took center stage. In that moment, I understood something clearly. We were no longer in an interview. We were stepping into the story behind every story he would tell.

The beginning explained everything that came after. Mississippi wasn't just where his life started. It was where the fight began.

Chapter 2: Born to Fight

"Mississippi taught me how to fight long before the Marines ever did. Down there, surviving was the first battle."— Lewis Cain.

"It started in Mississippi," he said.

Lewis was responding to my question, "What can you tell me about your childhood that shaped who you are today?"

He did not answer right away. His gaze drifted past the kitchen window for a moment, as if he had to locate the boy before he could speak for the man.

"I was born in Jackson, Mississippi," he began. "My mama had thirteen children. Nine boys and four girls. I was right in the middle of them."

A small smile touched his mouth, then faded as quickly as it came.

"When I was six years old, everything changed."

"I had rheumatic fever," he said quietly. "I couldn't walk. Couldn't move. My mama took me to the Children's Hospital in Jackson because she didn't know what else to do. I remember lying in that bed, watching other kids walk past my door, hearing them run up and down the hallway. Their footsteps sounded like freedom. And I couldn't even stand."

He glanced down at his hands, almost as if the memory put him back in that hospital bed.

"My mama came every day she could," he continued. "She'd rub my legs, pray over me, tell me I was gonna walk again. That woman didn't believe in quitting. She taught all of us that. She used to say, 'Lewis, the Lord didn't bring you here to leave you here.' And I believed her, because my mama never lied to us. Not once."

He paused, and his jaw tightened the way it does when a person is bracing for what comes next.

"One day I had this fire truck," he said. "Bright red. The only toy I had. And this white nurse came in and snatched it right out of my hand."

His voice dropped lower.

"I reached for it. I wasn't trying to hurt her. I just wanted my toy back. I hit her trying to grab it, and she looked straight at me and said, 'Even little black boys get lynched for touching a white woman.'"

My chest tightened at the bluntness of it. Six years old, in a hospital bed, hearing a threat that belonged to grown men and graveyards.

He continued, "Later that day, another nurse saw me crying," he said. "She didn't say nothing.

She just slipped the toy back into my bed when nobody was watching. Gave me a little smile. I never forgot her either."

He went quiet long enough for the story to settle.

Once the story settled, he continued again, "When I told my mama what that nurse said, she didn't waste a second. She came back that same night. Didn't argue with nobody. Didn't wait for permission. She looked at me and said, 'Get your things. You ain't staying here another minute.'"

Lewis shook his head as if he could still see it.

"She lifted me right out of that bed. I remember her arms under me. She was scared but steady. My mama always got steady when it was time to protect her children. She carried me through those hospital halls like she was carrying her whole world."

"She took me home, put me in my own bed, and she started praying," he said. "Every day. Sometimes twice a day. She'd rub my legs until her hands hurt. She'd say, 'You gone walk again, Lewis. The Lord didn't give you legs for decoration.'"

A faint smile returned, softer this time.

"She had a way of praying like she wasn't asking God," he said. "She was reminding Him."

"She kept at it," Lewis continued. "Day after

day. And slowly, real slow, I started to move again. First, a twitch. Then a little lift. Then one day she held me under my arms and said, 'Come on, put some weight on them legs.' And I did."

"My mama taught me how to walk all over again," he said. "Inch by inch. Step by step. She wouldn't let me quit. She didn't let any of us quit."

He leaned back slightly and blinked like he had to return from the past before he could keep going.

"That woman," he said, almost under his breath, "she saved my life more than once."

Then his face eased into something warmer, the kind of tenderness that shows up when hardship is remembered alongside love.

"You know, Mississippi was cold in the winter," he said. "Not just chilly. The kind of cold that gets in your bones. Our house was made of thin boards. My mama built it with her own hands, but the wind still came right through the cracks. You could see your breath in the air sometimes."

He recalled, "One night I was shivering so bad my teeth were clicking. I remember pulling my knees up to my chest, trying to warm myself. My older brother saw me."

Lewis's voice softened.

"He took off the only coat he owned and

wrapped it around me. He said, 'You're younger. You need it more.'"

He looked down at the table for a beat, then back up.

"That coat was rough against my skin, but it felt like love," he said. "That is what we had back then. Nothing fancy. But we had each other."

After that, his expression shifted into something almost surprised, as if he had stumbled into a memory he did not expect to share.

"I used to have this dream," he said. "It started when I was still stuck in that bed, and it kept coming back night after night."

"In the dream, I was running through a forest filled with sunlight," he said. "Bright and golden, it reminded me of Heaven. I could feel the wind on my face, cool and clean. My legs moved strong and steady beneath me. My lungs pulled in air like they were made for it. The earth under my feet felt solid, carrying me forward."

"I ran without pain," he said. "No stiffness, no weakness, no fever. Just me running like I had never been sick a day in my life."

I felt my shoulders lower as he spoke, like his dream gave both of us room to breathe.

"I did not understand it at first," he said. "I

thought maybe it was my mind trying to comfort me. But the more it came, the more I felt like God was showing me something. Like He was telling me, 'This is who you are; this is who you will be again.'"

"That dream became hope," he said. "Real hope. I believed God was giving me a glimpse of what was coming. A promise that I would not stay sick forever. That one day I would walk and maybe even run again."

A gentle smile touched the corner of his mouth.

"And He was right," Lewis said. "Mama prayed. God answered. Those dreams carried me until my legs did."

"She prayed every morning," he continued. "Before the sun came up. Talking to God like He was sitting right there in the kitchen. 'Lord, thank You for waking me up. Cover my children. Heal my boy's body. Do what only You can do.'"

"She'd lay her hand on my chest," he said, "and the tremors would calm. She'd tell me stories about Joseph in the pit, Moses crossing the waters, men God lifted from broken places."

He looked at me then, steady, as if he wanted me to understand this part as much as the pain.

"She was teaching me how to survive before I knew I'd have to," he said.

After a moment, he continued, “I learned real young the world wasn’t safe for little black boys. Not in Mississippi.”

He lifted his eyes toward mine.

“You want to hear what happened next?”

“I do,” I said.

Lewis shifted in his chair and drew in a breath.

“I thought I was ready to live again,” he said. “My legs were still weak, but I wanted to feel normal. So I went to the movies.”

“In Mississippi, colored folks had to sit upstairs,” he explained. “Never downstairs with the whites. We were sent to the top balcony, where it was hotter and crowded, but it was the only way we were allowed to watch the picture.”

“When the movie was over, we walked down the stairs to leave,” Lewis said, “and I saw them before I even stepped outside.”

“Police,” he said. “They were standing out front with dogs, big German shepherds, muscles tight, teeth showing, like they had been waiting for us.”

“There had been civil rights demonstrations earlier that week,” he said. “Folks protesting. Marching. And the police didn’t care who they scared or who they punished.”

"They didn't ask us no questions," Lewis continued. "Didn't care that we were just kids leaving a movie. The minute we stepped out, they let those dogs go."

His voice stayed controlled, but the scene he described did not.

"I turned to run," he said, "but my legs weren't strong yet. I felt them shaking, and that is when I fell."

"I hit the ground hard," he said. "I could hear the dogs behind me, their breathing loud, their growls low and closing in. There was no time to think. I just knew that if they reached me, it would be over."

"So I crawled," he said. "My hands scraped raw, my knees burning as I dragged myself forward anyway I could. I kept telling my legs, *Move. Move now*. And somehow, they listened."

He drew a breath and continued, "I got back on my feet, pain tearing through me, but I did not care. I ran with everything I had. People were shouting, dogs were barking, and my heart was pounding so hard it drowned out every other sound."

He swallowed hard.

"I didn't stop until those dogs were far behind us."

"Mississippi didn't care if you were a child," he said. "Didn't care if you were innocent. If you were Black, you were a target."

He was quiet for a moment, then added, "What steadied me was praying for myself. I remembered that dream God had given me, the one where I was running without pain. I reminded Him of it right there; told Him, 'You showed me I would walk again. You did not bring me this far to leave me now.' Holding on to that promise, that is how I made it out."

I watched his hands as they rested on the table.

"Lewis," I asked, "do you need a break?"

He looked up at me and nodded slightly, like he appreciated the question more than he wanted to admit.

"That one still gets me," he said.

"We can pause," I told him. "Whatever you need."

He shook his head. "No. I am alright. Talking about it brings it back, but I lived through it then. I can keep going."

"Okay," I said quietly.

Lewis exhaled, adjusted in his seat, and continued.

"There's something else," he said. "Something I saw when I got older."

"I was a young man then," he said. "Old enough to know better than to stand too close when trouble walked into a room."

"I remember being at a diner," he said. "My cousin was sitting at the counter. He had a glass of beer in his hand. Just a working man taking a break after a long day. The place smelled like grease and tobacco. People were eating, talking, minding their business."

He paused.

"And then the door opened."

"A white police officer walked in," Lewis said. "Heavy boots. Hard stare. He scanned the room and then locked his eyes on my cousin."

He said, "'Come here, boy.'"

Lewis's voice tightened on the word.

"Boy was never anybody's name," he said. "It was a way to strip you down. A way to remind you of the place they thought you belonged."

"My cousin lifted his beer," Lewis said, his voice low and steady. "He was calm. Worn out. Not disrespectful. Just a man trying to catch his breath at the end of a long day. He looked at the officer and said, 'What's the problem, officer? I'm just having a drink after work."

He paused, the memory tightening his jaw.

"The whole diner froze," he continued. "Forks stopped midair. Conversations died. You could feel everyone holding their breath."

His eyes darkened.

"Before my cousin could even stand, they were on him. Fists crashing down. Boots slamming into him. They beat him until his body collapsed onto the floor, like it had no strength left to hold itself up. Nobody moved. Nobody spoke. Then they dragged him outside and kept beating him in the dirt, like the world had decided his life didn't matter."

He swallowed.

"And then I heard it," he said. "A car engine starting."

"They ran over him," Lewis said. "Once. Then again."

"When they were done, they threw chains on him," he said. "Dragged him away like his life meant nothing at all."

"I wanted to run in," Lewis said. "Every part of me wanted to fight them, to step in and stop it somehow." He paused, then shook his head. "I could see where it was headed even before it happened. I knew how fast a moment like that becomes fatal, how quickly a single step forward can cost you your life." His voice lowered. "So I stood there, making a choice I hated, holding myself still while everything in me wanted to move, knowing that if I did, I would be the next body on the ground."

"That kind of thing changes you," he said. "It teaches you that in Mississippi, you could be minding your own business and still be marked for destruction."

He looked at me, and the steadiness in his eyes made the words heavier.

"Mississippi taught me about danger long before the Marine Corps ever did," Lewis said. "It taught me what it feels like to be watched. To be hunted."

He sat with that truth for a moment, then went on. "You do not grow up there without learning how to stay alert. How to read a room. Knowing when to move and when to stand still might save your life. I did not know the word for it back then, but that was training."

He lifted his eyes. "I learned how to fight before I ever put on a uniform. Mississippi made sure of that."

For the first time, the pieces aligned. The

sickness. The dogs. The threats. The restraint. The instinct to survive. This chapter was never just about where he came from. It was about how he was shaped. Mississippi was not simply a place he lived. It was the proving ground that forged him.

"By the time I joined the Marines," he said quietly, "I already knew what war felt like. I just thought maybe this time I would get to choose the battlefield."

He stopped there, allowing that truth to settle between us.

"I told myself I was ready," he continued. "I told myself nothing could be worse than what I had already survived."

He looked up then, steady. Certain.

"I was wrong."

He did not need to say the name right away. It hovered in the space between us, inevitable.

Vietnam.

I let the silence breathe before answering. "When you are ready," I said, "we can go there."

He nodded once.

And even seated at a kitchen table, it felt like we had crossed a line. Mississippi was behind us now, but the training it gave him would soon be tested in a place where survival meant something far more unforgiving.

Vietnam was not just another chapter.

It was the war Mississippi had been preparing him for all along.

Chapter 3: Death Lurks in the Elephant Grass

"Still waters run deep, boy. Quiet don't always mean safe."—Mary Lindsay Cain.

The kitchen quieted as Lewis settled back in his chair. Mississippi was still in the edges of his voice, but his eyes had already moved on, fixed on something farther away. Something he had not outrun.

When he spoke again, the words came steady.

"After everything that happened down there, I wanted to be strong," he said. "Strong enough to never run from dogs again. Strong enough to never feel helpless."

He paused, then said it plainly.

"So I joined the Marines."

I nodded. "Tell me about that."

Lewis leaned back as if his memory rose like heat off pavement.

"I went to boot camp in San Diego. Marine Corps Recruit Depot," he said. "First day, they looked at me and saw how skinny I was. They didn't know anything about my past. Didn't know I had spent years not walking. They just saw a kid who looked weak."

His jaw tightened, not with anger, but with something earned.

"I struggled. Bad. Especially with the running," he said. "My legs were still coming back from that bed. Every lap, I fell behind. I stumbled. Some days I hit that ground, and it felt like I was right back in Mississippi all over again."

He shook his head once, like he could still hear them.

"Those drill instructors yelled like the world was ending. But I was used to being pushed. I had been fighting to stand up since I was six."

He looked at me for a moment, then continued.

"One morning, after weeks of falling behind, something changed. I don't know if it was the training finally catching up to me or if my body just decided it was done being afraid. But I started running. Real running. And before I knew it, I passed the whole platoon."

He let the silence hold that image.

"That was the day I became a Marine," he said. "Not when they gave me a uniform. Not when they shaved my head. It was when I realized I could outrun what tried to break me."

Boot camp ended quickly. History was moving faster than any of them were ready for.

"President Johnson was sending troops fast," Lewis said. "Too fast. Before I knew it, I was on a ship headed to Vietnam in March of 1965. First ground troops."

His voice dropped slightly, not dramatic, just factual in a way that made it heavier.

"We didn't know what was waiting."

He stared past the kitchen wall.

"The heat hit you like a wall," he said. "You could smell the jungle before you saw it. And everything felt dangerous. Everything could kill you."

He breathed out and shifted, as if steadying himself.

"That's where I met the men who became my brothers."

He said Henry's name first.

"Henry was one of the first. He was from down South like me. Strong, steady. He carried the ammunition for the M60 and heavy cans across his shoulders on every mission. That job would break most men. Henry never complained. He always found something to laugh about, even when there was nothing funny."

Lewis's mouth lifted for half a second, then the expression went quiet again.

"His laugh kept a lot of us going."

Then he said Reed.

"Reed came not long after," Lewis said. "Quiet at first, but loyal. He watched everything. Stayed ready. Moved like he already knew what direction danger was coming from; he was one of the best riflemen we had. When you were walking behind Reed, you felt safer. You just didn't say it out loud."

He rested his hands on the table, fingers still.

"In Vietnam, it didn't matter where you came from," he said. "It didn't matter what you were back home. What mattered was trust. Who would pull you out of the mud when bullets started flying? Who would stay close when the jungle closed in?"

He looked down briefly, then back up.

"We didn't choose each other. The war chose us. And once we hit that first ambush together, that was it. We became brothers. Not by blood, by survival."

He paused.

"These were the men I trusted with my life," he said. "And when one of us fell, all of us felt it."

His eyes narrowed, not from emotion performed, but from memory tightening.

"When Henry died," he said, "everything changed."

The jungle rose around his words like a curtain coming down as Cain vividly recalled what happened that night.

Night lay over everything like a heavy cloth, wet and close, the kind of darkness that pressed against skin and crept down collars and into boots. A monsoon poured without mercy, not in drops but in sheets that stung cheeks and knuckles. The rain hammered leaves until the sound became a wall, swallowing everything else.

The air smelled of rot and sap and smoke left behind from the day's fighting. Mud splashed up Lewis Cain's legs, cold and thick enough to pull at his boots. He tasted metal when muck hit his lips. Grit worked its way under his nails and into the seams of his gloves.

Elephant grass towered over them, taller than helmets, sometimes taller than a man. From a distance, it looked like soft green waves. Up close, it was something else. Thick blades. Sharp edges. When wind ran through it, it hissed like a living thing. When it brushed your skin, it scraped just enough to sting, leaving thin lines that burned under rainwater. It moved against faces and necks and hands, as if the jungle were testing who belonged there.

The men moved cautiously through the darkness because the ground could bite. Beneath the surface, sharpened bamboo stakes waited inside hidden pits. Tripwires lay almost invisible, stretched across roots thin as hair. A wrong step meant torn flesh, infection, or death.

Lewis walked near the middle of the patrol. In front, the point man scanned the ground for lines or angles that did not belong to nature. Behind him, the M60 team advanced with the slow discipline of men who understood haste could get everyone killed. Farther back, the radio operator carried the PRC 25, the long antenna swaying above the grass like a fishing pole. Static hummed through the handset; the faint hiss served as a reminder that nothing in that jungle was fully silent.

Someone cursed under his breath when a branch snapped. Another Marine tugged his poncho tight. Sweat, fear, and wet canvas clung to them like a second skin.

Lightning cracked overhead, flashing white for a heartbeat, long enough to show how little they truly saw. Then, the darkness returned and swallowed them again.

Lewis took careful steps, muscles trembling under the weight of his pack. They had been trudging for hours, and still they kept moving.

Somewhere inside the storm, he heard his mother's voice the way you hear a remembered hymn.

Still waters run deep, boy. And quiet don't always mean safe.

Henry was near the front, weighted down by the machine gunners' ammunition. He had been joking earlier, trying to cut through the thick air of danger.

"You can tell my mama I carried enough weight out here to make me ten inches shorter," Henry had said before they left camp. "If I come home with my spine shaped like a question mark, y'all know what did it."

Lewis remembered laughing. Then, he remembered how quickly laughter disappears in war.

The rain intensified. The ground turned slick beneath their boots. They reached a swollen river, brown and churning under the storm, running fast enough to drag a log sideways.

The squad prepared to cross.

Henry stepped forward first.

"I got it," he said, forcing a grin. "Machine guns need to eat too."

Lewis opened his mouth to tell him to wait, but the river took Henry before the warning could form.

A foot slipped on a submerged rock. The current snatched him like something hungry. Henry's body spun as water dragged him sideways. His eyes widened. He clutched the ammunition and his rifle as if holding on to both might anchor him.

But the river only pulled harder.

"Henry," Lewis shouted, voice cracking against the roar of the storm.

Henry surfaced once, twice, fighting the current with everything in him. Then a wave swallowed him. For a moment, it looked like he had found footing.

Then, he disappeared.

The river gave no mercy.

Lewis dove toward the bank, skidding through wet soil, reaching for anything. He shouted Henry's name until his throat burned. The men fanned out along the shore, scanning for movement, but the rain blurred everything. The water carried Henry farther and farther away.

Days later, when they retrieved Henry's body, Lewis said the world tilted.

Henry's rifle was still clutched in his hands.

And his eyes were gone.

The crabs had taken them. Empty sockets stared up from the river's work like a warning.

Lewis dropped to his knees. Reed's hand landed on his shoulder, solid and wordless. There were moments in war when language was useless. This was one of them.

That night, under a weak lantern, Lewis and Reed sat in silence. No jokes. No toothpick at the corner of Reed's mouth. Just the sound of canvas breathing with the wind.

Lewis rubbed his thumb over his knuckle as he spoke, a small motion that never left him when a memory cut too deep.

"After we found Henry," he said, "Reed leaned over and whispered, 'We stick together now. No matter what.'"

Lewis looked at me as he repeated it, like he was still hearing the voice in that tent.

"And I told him, 'We come home together.' That's what I said. We meant it."

He nodded once, as if confirming it to himself.

"We clasped hands right there in the dark. Not for show. Not for comfort. A vow."

He swallowed.

"But the jungle had other plans."

He sat quietly after that. The only sound was the room around us doing what rooms do, ordinary and unaware. His hands were folded on the table, thumb still rubbing that same knuckle.

After a moment, I asked, "Lewis, what happened next?"

He inhaled slowly.

"After Henry," he said, "we had to go right back out there."

Hours later, they were back on patrol, although no one wanted to be.

"There was no laughing," Lewis said. "No joking around. Every man walked like he had a ghost behind him."

The jungle felt different after losing a man. Even the leaves seemed to watch them.

The rain returned in scattered drops, tapping the tops of helmets. Then the sky opened again, a monsoon in sheets, heavy enough to blur the path ahead.

"The air was thick," Lewis said. "Like breathing through a wet cloth. And the mud pulled at our boots like it wanted us to sink."

No one spoke Henry's name, but he was there anyway. In the river behind them. In the empty space where he should have been walking. In the quiet where his laughter used to land.

"Reed wasn't the same," Lewis said.

Reed walked stiffly, jaw clenched, shoulders tight as if bracing for impact. Lewis described how Reed had slept with his hand resting on a grenade the night before, flinching at every distant sound.

"He looked older after one night," Lewis said. "War takes time off you."

Somewhere behind them, a Marine whispered, "Feels wrong to be out here."

The rain almost stole the words.

Another voice answered, "Feels like everything is watching us in this grass."

Lewis nodded as if he could still hear it.

"That grass grew taller than your helmet," he said. "Sometimes taller than a man. You could be two feet from somebody and not see them. And when it brushed you, it wasn't soft. It pressed your face, your neck, your hands. Like fingers trying to find a weak spot."

The jungle was too quiet.

No birds.

No frogs.

Not even the rustle of small creatures hiding from the rain.

The silence felt alive, coiled and listening.

The point man raised a fist, signaling them to slow.

Bodies tightened. Weapons shifted. Eyes strained through rain.

The M60 team adjusted their footing. The radio operator lifted the handset closer.

Lewis said, "Every step felt like stepping into a trap."

Reed glanced back, eyes meeting Lewis's for a brief flicker. No speech. Just a look that said, 'Stay close.'

Lewis nodded.

They stepped forward.

Thunder rolled across the canopy. Lightning cracked the sky.

And then something shifted in the darkness.

A rustle.

A flicker.

A shadow not shaped like a tree.

The patrol froze.

For one suspended breath, the jungle held still.

Then, it exploded.

Gunfire tore through the night. Muzzle flashes lit the darkness in jagged bursts. Tracer rounds carved bright arcs through the monsoon. Shouting rose and fell under the roar of rain.

Lewis dropped flat. His chest hit the mud hard enough to knock the breath out of him. His helmet cracked against a rock. Mud filled his mouth, gritty and metallic. He rolled, coughing, wiping his face with the back of his glove.

When he looked up through rain and blur, Reed was sprinting for a tree, trying to reach better cover. His boots slid on soaked ground, each step desperate, breath tearing from his chest as the world cracked open around them.

He was almost there when a single shot split the storm.

Reed's body snapped forward as if yanked by an invisible hook. His helmet spun off and vanished into the mud. Blood arced into the rain and disappeared as quickly as it came.

Reed hit his knees. His hands flew to his throat.

Lewis's voice dropped at my table, as if the moment still demanded quiet.

"No," he said.

Then, he kept going.

Lewis crawled toward him on hands and knees, bullets screaming overhead, mud sucking at his palms. Reed's uniform darkened fast, rain diluting blood but not stopping it. Reed tried to speak, but only a choking sound came out.

Lewis pressed both hands to the wound.

Blood spilled through his fingers, hot and slick.

Reed's eyes were wide, fixed on Lewis like he was trying to stay tethered to something human.

"Stay with me," Lewis begged. "Please. Stay with me."

Reed's legs twitched once, a small involuntary jolt. His breathing hitched, fought, then faltered. A final bubble of blood rose and burst against the rain.

Then, he went still.

Thin streams of red ran down his chin and disappeared into the mud.

Lewis swallowed hard.

"His eyes just stayed open," he said.

The jungle did not pause out of respect. Bullets still cracked through trees. Rain kept roaring. Darkness pressed in around them.

"Somebody call the medevac," Lewis yelled into the storm.

The radio crackled back with an answer colder than the downpour.

The helicopter would not fly into incoming fire.

Reed would not be saved.

Lewis bowed over his friend's body. For a moment, the noise around him blurred into something distant, like the war had moved away and left him alone with the vow he could not keep.

"First Henry," he whispered. "Then, Reed. Two brothers gone. Just like that."

Across the table, Lewis's hands trembled. Not dramatic. Not for effect. The movement was small, persistent, like his body still carried the vibration of that night.

I leaned forward, careful with my voice. "Lewis, are you okay?"

He nodded once.

"I'm fine," he said, though he didn't sound fine. "I just remember it like it was yesterday."

He drew in a slow breath, as if choosing whether to open the next door.

“That night,” he continued, “I couldn’t sleep. Every time I closed my eyes, I saw them: Henry’s empty sockets. Reed falling in the rain. I could hear the jungle around me like it was whispering.”

His throat tightened. He swallowed hard.

“I thought the next death would be mine.”

He looked past me, not at the room, not at the present.

“That jungle was alive, Dr. Kara,” he said. “And it wanted all of us.”

He paused.

“And what happened next,” he said quietly, “proved it.”

Chapter 4: A Storm of Wings and Death

"The jungle was talking to me that night; it remembered I was the one still alive, and I knew it was coming for me next."—Lewis Cain.

Lewis stared down at the kitchen table as though the wood itself could remember the things he was trying to describe. His hands had begun to tremble again, the way they did whenever a memory rose too quickly for him to brace against. I had grown familiar with that tremor during our interviews, but this time it carried a deeper shiver, something older and heavier than the others.

"It was the night after Reed died," he said quietly. "Sleep just... would not come."

He lifted his eyes for a moment, and the kitchen disappeared behind them. I knew he was no longer sitting across from me. He was back in the jungle, lying on a thin poncho liner, hidden under a canvas tent that did nothing to keep the cold fear from settling into his bones.

"It was dark in that tent," he said. "This wasn't ordinary darkness. This one pressed in, heavy and deliberate, as if it carried a purpose. Every time I closed my eyes, I saw Henry's face, or what was left of it. And then Reed. Falling in the rain. Holding his throat. I heard him choking even when the jungle went quiet."

I watched his throat tighten as he swallowed. He took slow breaths as if pacing his body back into the present.

"I kept seeing the river taking Henry," he whispered. "And Reed... the way his helmet rolled away. The way his eyes stayed open. I could not stop seeing it."

He closed his eyes briefly, and for a moment I thought he might stop. I almost told him he could. But he continued before I could speak.

"I tried turning on my side. I tried lying on my back. It was no use. When I finally closed my eyes, the darkness behind them was worse than the darkness around me. Every sound made my heart jump. A twig outside. A drop of rain hitting the canvas. My own breath. Everything sounded like danger."

He rubbed a thumb along the rim of his coffee cup. The muscles in his jaw flexed as if the memory itself made breathing difficult.

"Dr. Kara, that jungle was talking to me that night," he said. "I felt it. Like it remembered I was the one still alive. Like it was waiting to see what it would take from me next."

His stomach tightened as he spoke. I could see it in the way he pressed his palm against his abdomen. His breath grew shallow. His fingers curled slightly as if holding onto something invisible.

"You ever feel like the ground is breathing underneath you?" he asked softly. "Like you are lying on something that is awake?"

I nodded gently, not because I knew that feeling, but because I wanted him to keep going in whatever way felt safest for him.

He continued. "I lay there shaking. My hands would not be still. My legs kept twitching like they were bracing to run even though there was nowhere to go. I kept counting the breaths between each sound outside. My mind was racing, but my body felt frozen. I was twenty years old, and I felt like a child again. Scared to close my eyes. Scared to stay awake. Scared of everything."

Lewis continued, "I did not know what to call it back then. I just thought something was wrong with me. I thought maybe the jungle had finally gotten inside my head."

A long silence fell between us. I watched his hands tremble in a way he tried to hide by folding them into each other. His shoulders rose and fell slowly, controlled, as if he were determined not to let the memory overpower his present.

"Lewis," I said softly, "are you alright?"

He nodded once, but his voice was not steady when he answered. "I just remember it like it was happening right now."

And I could see it.
The boy he had once been.
The Marine he had become.
The man sitting in front of me trying to hold all those versions of himself together.

In that moment, I understood something deeply.
The war had ended for everyone else.
But for Lewis, that night had never let him go.

"And that was just the night," he said, his voice falling to a whisper. "Morning came. And the next thing that happened... nobody should have survived."

He paused; the air in his kitchen felt still, as if it, too, were holding its breath.

"Dr. Kara," he said, "that next day... is when the jungle truly came for me."

Lewis sat back in his chair, his eyes distant for a moment as the memory took him back. His hands trembled slightly, the same as they had when he was recounting the deaths of Henry and Reed. But this time, the tremor was deeper, more ingrained.

"It was the next morning," he began, his voice low but steady. "Dawn cracked the sky, but I don't think any of us had slept. The storm had come and gone, but the air… it was still thick with it, with the loss of Henry and Reed, and that damn jungle that felt like it was closing in on us."

He paused, swallowing hard, as if he could feel that day still pressed on him.

"We were all dehydrated, you know? The kind of thirst that gnaws at your insides until it becomes a physical pain. And our water supply had gone down to nothing. We'd been out there too long without a resupply, and I could feel it in my body, that dry ache in my throat. The air felt like you were sucking dust instead of oxygen."

He leaned forward, eyes locking onto mine with a sudden sharpness. "Then, we saw the helicopter. A bird coming from far off, cutting through the fog. It was bringing us water. I don't think any of us cared about anything else in that moment, just the promise of liquid running down our throats."

"But someone had to get it," he continued, the tension returning to his voice. "The crates were dropped about half a mile away, but the jungle wasn't safe. Too much could happen in that short distance. Everyone was looking around, but nobody moved. Nobody wanted to be the one to walk out there and risk it."

Lewis looked down at his hands again, his fingers clenched tightly as if he were still holding onto that rifle from that day.

"I volunteered," he said, his voice softening. "I didn't even think about it. I just... I needed to move. Needed to do something. Sitting still, waiting for another ambush to hit, for another life to be taken... I couldn't do it. I couldn't feel that helplessness again, not after Henry. Not after Reed."

I could see the pain in his eyes as he relived that moment. I understood then that it wasn't just the water he was after. It was control. It was action. A desperate need to fight against the helplessness that had been swallowing him whole since the river took Henry and the bullets took Reed.

"But I didn't know then what I was walking into," he murmured, his voice barely above a whisper. "I didn't know that my next step would take me into something far worse than anything I had already faced."

And as he sat there, eyes glazed over with that memory, I knew something terrible was yet to come.

Lewis shifted in his chair, his eyes darkening as he began to speak, and I could feel the significance of what he was about to say in the quiet before his words. The storm outside the kitchen window seemed to mirror the tension building in the room. He was quiet for a moment, gathering his thoughts, but I could tell he was already there, back in that jungle.

"I remember the air," he started, his voice even, but heavy. "The way it felt that day, like everything was alive. The humidity, the rain, the sweat... it all became one thing, thick and suffocating, like the jungle itself was wrapping its hands around your neck." He paused, eyes far away, and then slowly looked up at me. "We had to get water from a drop point. You know, the usual. The helicopter had come in, dropped the crates, and I was one of the men who had to retrieve them. We didn't have much choice."

He leaned forward, his voice lowering. "I was walking through what I thought was elephant grass. The tall, thick stuff that's everywhere in Vietnam. You can barely see past it. It sways with the wind, cuts your skin, and you move through it like you're wading into a deep ocean, each step dragging. But something was wrong. The grass— it moved differently. The air, too. It felt like it was vibrating, humming beneath my feet."

Lewis swallowed hard, his hands tightening into fists at his sides. I could see the tension in his jaw, the way his body instinctively tensed as if he were back there, stepping into something he couldn't control.

"Then, it happened," he continued, his voice shaking slightly. "At first, it was just the buzzing. Soft, almost like a faraway hum. But then it grew. It came from everywhere at once. The air was alive with it. I knew immediately. I knew what was coming."

His face twisted with pain as the memory overtook him.

"And then they were there," he said, his breath quickening. "Bees. Hundreds of them. No, more like six hundred. A swarm, driven by the helicopter blades, pushed toward me like they knew exactly where I was. It wasn't just a few, or even a dozen; it was a wall of them. They hit me all at once. Covering my face. My mouth. My eyes. Everywhere. I couldn't breathe, couldn't move fast enough to get away. They were inside my mouth, crawling into my nose. One got stuck under my eyelid."

The room felt smaller as he spoke, the space between us filling with what he was remembering.

"I screamed," he said, "not from fear, but from pain. When I hit the ground, it was so hard it felt like my spine shattered on impact. I started throwing my body back and forth in the dirt, rolling, thrashing, trying to knock the bees off me, hoping they would finally stop. The pain in my back kept building with every violent movement until my legs locked up. I couldn't feel them anymore. I couldn't move. It was like my body shut itself down, like paralysis set in all at once. And then, everything went dark."

He glanced down at his legs for a moment, then lifted his eyes back to me.

"It shut my body down one sting at a time," he said. "My breathing went first. Then my sight. They were everywhere, in my face, in my mouth, and the helicopter blades were pushing them straight into me, driving them deeper, faster."

He paused and took a slow breath, as if his body still remembered what it took to get air.

"They told me later I'd taken close to five hundred stings," he said. "They said my heart stopped. They loaded me onto the same helicopter that brought the water in, and that's the last thing I remember."

When he opened his eyes again, Vietnam was gone.

"I woke up at Clark Air Force Base in the Philippines," he said.

The doctors told him the venom alone should have killed him. That once his heart stopped, there should have been no coming back from that.

"They kept saying I wasn't supposed to be conscious," he said. "Wasn't supposed to be walking. Wasn't supposed to be here."

He rubbed his hands together slowly as he spoke, steady but deliberate.

"I was huge," he said. "Swollen from the venom. I barely recognized my own body. And as soon as the poison cleared, they sent me back to Vietnam. No rest. No recovery. Nothing."

He shook his head once.

"I was still trying to breathe right and still shaking. My back pain was worse, I was stiff, and my body ached all over. I could barely walk, but none of that mattered. Surviving didn't mean you were finished," he said. "It just meant you were still usable." For a beat, neither of us spoke.

The room went still after that.

I have always been terrified of bees, and as I sat across from Lewis, listening to him describe the swarm, I felt my body react. My chest felt tight. My palms were damp. His breathing had changed, too. I could hear it, slow but strained, as if the memory had pulled us both into the same physical space.

I did not want to interrupt him or soften what he had lived through. Nothing about that moment deserved to be rushed or explained away. So we sat there quietly, letting the refrigerator hum fill the space between us, the only sound in the kitchen.

After a few moments, I asked gently, "Do you want to move on to something else for a bit?"

He nodded almost immediately. "Yeah," he said. "That would be good."

I took a breath and shifted us forward.

"What was it like," I asked him, "being a Black man among the first ground troops in Vietnam?"

His voice shifted, steadier now.

"I had never seen Black people like that before," he said. "Not together like that."

He explained how Black soldiers quickly found one another, how they protected one another, and how the phrase "soul brother" meant something real.

"We were tight," he said. "One for all. All for one. We didn't fight each other. We watched each other's backs."

In Vietnam, he said, survival erased the lines that had followed them since birth. Brotherhood was not symbolic. It was necessary.

"And that's why it surprised me when I came home," he added.

He paused, then continued.

"I never really understood why we were there," he said. "It didn't make sense. Back home, Black churches were being bombed. Children were being killed in Sunday school. Civil rights workers were being beaten, jailed, and murdered. Alabama was still burning with hatred. Mississippi was still dangerous. The Civil Rights Act had not even passed yet. And I was in Vietnam."

He shook his head slowly.

"They said we were fighting for freedom. But Black people didn't have freedom at home."

He spoke then of Muhammad Ali, not as a celebrity but as a man who saw the contradiction clearly. Ali had said he would not fight a war for a country that denied him basic rights; that no Vietnamese man had ever called him a racial slur. That he would not travel thousands of miles to kill people who had never harmed him, while Black Americans were being lynched and terrorized on American soil.

"Muhammad Ali was right," Lewis said. "He said it didn't make sense to fight for something overseas that we didn't have at home."

He spoke next of Malcolm X, whose words had reached young Black men like him even across oceans. Malcolm had argued that Black people were being used. Sent to die for a government that refused to protect them. Drafted into wars while their neighborhoods were neglected, their churches burned, their lives treated as disposable.

"Malcolm said we were being manipulated," Lewis said.

He nodded once.

"When he said by any means necessary, I understood it later. He wasn't talking about violence. He was talking about survival. About dignity. About not letting somebody keep using you."

Lewis leaned back slightly.

"We were fighting in Vietnam," he said, "while still fighting to be treated like men back home."

He was quiet for a moment.

"I just wish things had been different," he said.

Somewhere in his memory, the helicopter was already descending. The rotors were still turning. Orders were already in motion.

Vietnam was waiting.

And Lewis Cain was being sent back into it.

Chapter 5: Delayed Combat Fatigue

"Sometimes the war does not hit you when you're in it. Sometimes it waits. Sometimes it shows up later, when your body finally thinks it's safe to stop fighting."—Lewis Cain.

When I asked Lewis about Morocco, he did not answer right away. He leaned back just enough for the memory to settle, the way a man does when he knows the story is already standing in the room with him.

"That was after Vietnam," he said. "They sent me there like it was supposed to calm things down."

Morocco was not a battlefield. It was a reassignment, a holding place after combat, after the jungle, after everything that had tried to kill him. On paper, it was a quiet duty. No patrols cutting through elephant grass. No helicopters rattling the sky. No nights listening for movement that meant death.

But Lewis had learned long before Vietnam that quiet did not mean safe.

Men sat around in the evenings talking and drinking, filling the hours the way men do when the day has finally loosened its grip. The kind of time that leaves space for memories to surface and tempers to slip. Lewis said he was seated at a table when he heard the voice.

It was loud. Certain. Carrying a confidence that felt older than the room itself.

"I'm going to whoop all you boys."

Lewis told me it came from a white man from Alabama. He said you could hear the certainty in it, the kind that does not ask for permission or wait to see who might answer. A confidence built long before that night, shaped by Jim Crow laws, racist bloodlines, and generations of men who had been taught that calling Black men boy was not just acceptable, but powerful. It was a word meant to strip adulthood, dignity, and humanity in one breath. A word designed to remind you of where they believed you belonged.

Lewis said he knew that tone the moment it landed. He had heard it before. Mississippi rose up inside him without warning. Dogs straining against leashes. Police standing loose and ready. Black men dragged from porches and sidewalks for less than a word, for less than a look. None of it came as thoughts. It came as recognition.

Before Lewis could respond, the man crossed the space between them and struck him over the head with a bottle. Glass exploded. The sound cut through the room. And something inside Lewis, already stretched thin by war and memory, snapped into motion.

He said that was when everything he had carried finally collided. Mississippi. Vietnam. Bullets flying past his ears. Elephant grass slicing skin. Bees pouring into his face until his heart stopped. The jungle taking Henry. The rain swallowing Reed.

"I had already been through too much," he said quietly.

He stood up so fast the chair tipped back behind him. The switchblade was already in his hand before he realized it was there. Not drawn in thought. Not planned. Just there. The distance between them closed in an instant.

"I cut his throat," Lewis said. "From ear to ear."

He did not dramatize it. Sitting across from me at the table, Lewis lifted two fingers and drew them slowly from one side of his neck to the other, deliberate and exact. I watched the motion in stunned silence, my breath catching as I followed the line of his hand. There was no flourish. No exaggeration. Just a measured gesture, the kind a man makes when he is describing something that already happened and cannot be undone.

"The man went down, and when his hands went to his throat, I saw Reed," Lewis said. "Just for a second. Reed in the rain, holding his neck the same way. Blood was coming through his fingers. That jungle came right back to me. It wasn't something I thought about. It just showed up. Same motion. Same look in the eyes when a man knows he can't breathe. And then it was gone."

The room erupted with shouting, movement, and bodies surging all at once. Lewis said the sounds blurred together after that. Orders. Hands grabbing him. Weapons drawn. The feeling of being restrained returned so quickly that it felt rehearsed.

I sat there, unsure whether to ask if the man lived or died that night. Part of me wanted to know. Another part of me knew I should not. Lewis seemed to sense the question as he read my shocked expression. "He didn't die," he said, reassuring me. "But I was ready to let him."

I asked him what happened next.

"The military police were on me before the bottle hit the floor," Lewis said. "Hands on my arms. Orders coming from everywhere at once. I did not fight them."

He said someone later told him he was lucky the man lived. Lewis did not answer. He had heard versions of that sentence his whole life, always delivered as a warning, never as understanding.

"They talked about prison," he said. "Twenty years. That is what they said."

Instead of a cell or immediate charges, instead of the twenty years they had already mentioned, a doctor was brought in. Not a commander. Not an investigator. A doctor.

"That's when they said it," Lewis told me. "Delayed combat fatigue."

"That's what they called it," he said again, as if testing the phrase even now. "They said sometimes the war does not hit you when you're in it. Sometimes it waits. Sometimes it shows up later, when your body finally thinks it's safe to stop fighting."

He told me they explained it like this: the mind absorbs too much for too long, keeps moving because it has to, keeps surviving because there is no other option. Then one day, without warning, something small breaks the seal like a single word or a threat; a bottle raised in the air. And everything the body has been holding back comes out at once.

"They said my reaction wasn't about that man," Lewis said. "They said it was about everything before him."

Vietnam. Mississippi. Dogs. Guns. Orders

screamed in the dark. Friends dying in front of him. Being told to keep moving anyway.

"They said my system was overloaded," he told me. "That I snapped, not because I was violent, but because I had been holding it together too long."

That diagnosis changed the room.

The talk of prison quieted. The certainty of punishment loosened its grip. Instead of shackles and a cell, paperwork began moving in a different direction.

"They didn't say I was innocent," Lewis said. "They said I was injured."

Delayed combat fatigue meant he was no longer just a Marine who had crossed a line. He was a Marine whose mind had not yet come home from the war.

"It kept me out of prison," he said plainly. "That's the truth."

Within days, the decision was made. Lewis would not stay in Morocco. He would not face a court-martial. He would be sent for evaluation.

That was how he found himself on another plane.

"There were two MPs," he said, holding his hands a few inches apart. "One on my left. One on my right. They didn't talk to me. They just watched."

He shook his head once, a dry half smile crossing his face. "I felt like Hannibal Lecter or something. Like I was dangerous just for breathing."

No one spoke to him. No one explained where he was going. He was not under arrest, but he was not free either.

Lewis stayed still as the engines leveled out, his shoulders squared, his hands resting flat against his thighs. The two MPs remained silent, close enough that he could feel the heat from their uniforms. He did not look at them. He did not need to.

When the word came, it was quiet and final.

"Wiesbaden."

Nothing else followed it.

Lewis knew that name. Every Marine did. Wiesbaden was not a posting. It was a pause. A place where men went when the Corps needed to decide what they were going to do with what came back from war. It meant doctors instead of officers. Charts instead of charges. Observation instead of answers.

"They didn't say hospital," he told me. "But that's what it was."

He stared at the metal floor of the aircraft as it vibrated beneath his boots. Germany meant evaluation. It meant being watched. It meant someone was finally acknowledging that whatever had happened in Morocco did not start there.

"It meant they were trying to figure out if I was broken," he said. "Or dangerous. Or both."

No one explained the process to him. No one asked him what he remembered or what he felt. He was moved because movement was easier than conversation. He was being carried from one decision to the next without being invited into any of them.

When the plane descended, the air changed. Colder. Thinner. The kind of cold that cleared the head but stiffened the body. Lewis stepped down onto the tarmac with the same discipline that had carried him through patrols and ambushes. Straight back. Eyes forward. Waiting.

Wiesbaden stood ahead of him, clean and quiet and controlled.

"And that's when I knew," he said. "This wasn't about what I did in Morocco, but everything that came before it."

Germany did not end the war. It simply marked the next place where it would be examined.

Wiesbaden Army Hospital was quiet in a way that made Lewis uneasy.

No helicopters. No shouting. No boots hitting dirt. The air smelled clean, almost sharp, and everything echoed just enough to remind him that he was indoors, contained, watched. He said the stillness made his body feel loud.

"They put me in a small room," he told me. "Desk. Two chairs. The window cracked just enough to let the cold in."

The doctor came in with a folder tucked under his arm, with a German accent. Calm voice. The kind of calm that made Lewis feel like he was the only thing in the room that did not belong.

"He didn't ask me about Morocco right away," Lewis said. "He asked me about Vietnam."

The doctor asked where Lewis had been stationed. Who he served with. How long did he sleep at night? Whether loud sounds made him flinch. Whether he ever felt like he was still there, even when he wasn't.

Lewis answered slowly. Carefully. He had learned that words could move you in directions you did not choose.

Then, in the middle of the questions, something shifted.

Lewis stopped speaking.

“I was sitting there looking at the doctor,” he said, “and then Reed was there too.”

Not standing. Not dramatic. Just there.

Seated slightly to the side, close enough to be unmistakable. Helmet tilted back. Uniform soaked through the way it had been that day. One hand pressed against his throat.

Lewis said the room did not change. The desk stayed solid. The doctor kept talking. But Reed stayed.

“I could see his fingers,” Lewis said. “Red. Wet. Slipping.”

The doctor asked another question. Lewis did not hear it.

Reed looked at him, not accusing, not angry. Just present. Waiting. The same way he had looked when he was trying to breathe and could not.

Lewis said his chest tightened, not like panic, but like recognition.

“I knew I wasn’t in Vietnam,” he told me. “But my body didn’t.”

The doctor noticed the pause and asked Lewis to look at him and tell him where he had just been.

Lewis did not answer right away.

"I didn't want to tell him Reed was sitting there," he said. "I didn't want to give that thing a name."

The doctor followed his gaze anyway. Looked at the empty chair. Then, back at Lewis.

"That's when he closed the folder," Lewis said. "That's when he stopped asking questions."

They sat in silence for a moment that felt longer than it was. Then the doctor spoke, neither sharply nor softly. Just clear.

"You are still fighting," he said.

Lewis said that sentence landed harder than the bottle ever had.

The rest of the session moved quietly. Notes were made. Words were written down that Lewis did not see. When it was over, he was told he would be staying for observation. No charges. No prison. No explanations felt complete.

"They didn't say I was dangerous," Lewis said. "They said I was delayed."

Wiesbaden became a place where days blurred together. Meals. Walks. Short conversations. Long nights. Reed appeared more than once. Sometimes sitting and sometimes standing just out of reach.

Always silent.

"They weren't trying to punish me," Lewis said. "They were trying to understand something they had created."

When the orders finally came, Lewis was not surprised.

Another transfer. Another plane.

This time to Great Lakes. Closer to home, they said.

Lewis said Reed did not board that plane with him.

But the empty seat beside him stayed open all the same.

Great Lakes Naval Hospital sat close to the water, wide and low, built to feel steady. Lewis said he noticed the lake first. Gray. Flat. Endless in a way that felt different from the ocean and nothing like the jungle.

"It was supposed to be close to home," he said. "That's what they told me."

The processing started again.

More forms. More questions. Different doctors asking versions of the same things. Where he had been. What he remembered. What hurt. What did not? They checked his body carefully. His shoulder.

His back. His reflexes. They asked him to walk. To lift. To sit. To wait.

"They were trying to figure out where to put me," Lewis said. "Not how to fix me."

At Great Lakes, he was quieter. He had learned that answers could travel faster than he did. He gave them what they asked for and kept the rest close. Vietnam stayed mostly unspoken. Morocco was not mentioned. Wiesbaden existed in the file, not in conversation.

"They already had a name for it by then," he said. "So they didn't need the story."

The days moved slowly. Medical processing turned into observation, then into waiting. Lewis said the waiting was the hardest part.

"No jungle. No orders. Just time."

He slept in short stretches. Ate when told and walked when directed. He said he could feel something shifting in him, not healing, just settling into a shape he did not recognize yet.

Then one afternoon, a doctor told him he was being transferred again.

No explanation beyond that.

"They said North Carolina," Lewis said. "Camp Lejeune."

He knew the name. Every Marine did. A base. A place where you finished things. A place where paperwork closed out a life and opened another one, whether you were ready or not.

"They didn't say discharge right away," he said. "But I knew."

Another plane. Another seat. Another stretch of sky.

Lewis said he watched the land change beneath him as the plane moved south. The water gave way to trees. The air was growing thicker before he even touched the ground. By the time they landed, he understood something clearly. Vietnam had ended on paper, but his movement had not. While Camp Lejeune was his discharge station, it was not the end; it was the next place the war learned his name.

Camp Lejeune was different.

Lewis said his wife and son joined him there. They lived on base, close enough for family life to look almost normal from the outside. A house. A routine. Mornings that did not begin with gunfire or helicopters.

"I had my family with me," he said. "That mattered."

He did not offer many details about those days, and I did not press. What he shared was enough to understand the shape of it. He was present. He showed up. He tried to step into the role that came next.

But the evaluations continued.

Medical appointments. Behavioral assessments. Conversations that circled the same ground from different angles. Lewis said they asked about sleep. About anger. About focus. About memories that came uninvited.

"They didn't call it PTSD back then," he said. "But they knew something wasn't right."

At Camp Lejeune, the symptoms were no longer treated as temporary.

The nightmares followed him there. So did the jumpiness. The way his body reacted before his mind had time to decide whether something was dangerous. He said he startled easily. Stayed alert even in quiet rooms. Slept lightly, if at all.

"They wrote it down," he said. "Everything was getting written down."

His shoulder pain was documented again. The back injury from the fall during the bee attack followed him into these records too. Lewis said the pain was already settling in by then, the kind that does not fade but learns how to wait.

"They kept evaluating," he said. "Like they were

building a picture."

Camp Lejeune became the place where the pieces started to connect—not yet fully named or acknowledged. But no longer ignored.

"They weren't asking if I could go back to Vietnam anymore," Lewis said. "They were asking what was coming home with me."

It was there, during those final evaluations, that the war stopped being something that had happened to him and became something he carried. And when his discharge finally came, it did not feel like an ending.

By the time he reached the end of it all, nothing felt separate anymore. Mississippi bled into Vietnam. The threat behind a man's voice sounded like barking dogs and police commands. Gunfire snapped through memory the same way glass had shattered on the floor in Morocco. Elephant grass still cut at his legs. The bees still filled his mouth and lungs until his heart stopped. The river was still taking Henry. Reed was still on his knees in the rain, hands at his throat. None of it stayed where it belonged. It all lived in the same place now.

He said leaving did not feel like freedom. It felt like a release without instructions. "I left Camp Lejeune," he said. "But nothing left me."

Chapter 6: Dreams in Uniform Denied

"I didn't lose my place. I followed the rules. They just decided the rules didn't apply to me."—Lewis Cain.

Lewis said he remembered the day he left Camp Lejeune because it felt unfinished.

It was May of 1968. He had his discharge papers in hand, a small bag over his shoulder, and the sense that something important had ended without ever being named. There was no ceremony, no closing words, no acknowledgment that marked the distance between who he had been and what he had survived. The base did not pause as he walked away from it. It stood exactly as it always had, orderly and unmoved, as if nothing had passed through him at all.

"I walked out with my papers," he told me. "That was it."

He did not describe relief. What he described was uncertainty, the kind that settles in when structure disappears, and no one tells you what comes next. Camp Lejeune had been the place where his body was examined and reexamined, where doctors documented injuries without ever fully naming the cost, where questions were asked but answers stayed partial. It was the final stop in a long chain of movement that had started in Vietnam and carried him across oceans and hospitals before depositing him back on American soil.

When he left, he believed he was stepping into something better.

“At that point, I still believed in the system,” he said. “I thought if you did what you were supposed to do, it counted for something.”

He had worn the uniform through places that demanded everything a man had to give. He had survived combat, loss, and the long aftermath that followed men home even when the war itself ended. He assumed that service would translate into opportunity, that discipline would still matter once the uniform came off, and that being a veteran meant the country would meet him halfway.

Lewis was twenty-three years old. He was trained, disciplined, and accustomed to structure. What civilian life offered instead was silence. No orders. No direction. Just the expectation that he would figure it out on his own.

He stayed in North Carolina for a short time after his discharge, trying to decide what the next step should be. He looked for work. He listened for guidance. He waited for something that never quite arrived.

"I needed a job," he said. "I needed something steady. I needed purpose."

He believed his record would open doors. He believed his service would speak for him. He believed the years he had given would carry credibility beyond the military gates.

"I thought being a veteran meant I wouldn't have to beg," he said.

What he encountered instead was a quiet indifference. There was no hostility, no outright rejection, just the steady closing of doors that never fully opened in the first place. Employers listened, thanked him for his service, and moved on. The uniform that had once commanded instant authority now existed only as a line on a form.

Lewis said he began to feel the impact of that shift in his body. The sleeplessness followed him. Nights stretched long and restless, his mind alert even when his body was exhausted. He would wake suddenly, unsure why, his heart already racing as if responding to a threat he could not see.

"I couldn't shut it off," he said. "I didn't know how."

He needed work not just for income, but also for structure. Something that would give his days shape again. Something that would quiet the constant readiness that had nowhere to go. It was during that stretch of uncertainty that he heard about the Comprehensive Employment and Training Act (CETA) program in Lakeside, WI, a federal initiative designed to help returning veterans transition into civilian employment.

"That sounded like what it was supposed to be for," he said. "Guys like me."

Lakeside was not home, but it represented possibility. He knew the city had industry. He knew it had government jobs. He believed a program created for veterans would recognize what he carried, even if he did not yet have language for it himself. So he packed what little he had and headed north.

When he arrived in Lakeside, the city felt colder than North Carolina in more ways than one. The buildings were taller. The pace was different. But Lewis did not come looking for comfort. He came looking for a place to stand.

The CETA program became his entry point. On paper, it was exactly what he needed. Training. Placement. A pathway into steady employment. It offered structure again, and Lewis responded to that immediately.

"I did what they asked," he said. "I showed up. I followed instructions."

Through the program, he was placed with

Lakeside County and eventually entered the Sheriff's Department. The uniform returned to his life in a different form, but the feeling was familiar. Order. Rank. Expectations. He knew how to function inside systems like that.

"At first, I thought I'd found my place," he said.

He believed his service record would matter there. He believed the veteran preference he had been promised would apply. He believed that promotions would come according to the rules he had already learned to live by.

Lewis did not yet know that another kind of war had already begun, one that did not announce itself with gunfire or orders, but with silence, delay, and decisions made behind closed doors. And it would take years before he understood just how much had been decided without him.

Lewis said the work itself came easily to him. He understood hierarchy; he understood procedure. He knew how to move inside an institution without drawing attention to himself. The Sheriff's Department felt familiar in that way, another uniform, another chain of command, another place where rules mattered more than feelings.

"I knew how to do the job," he said. "I wasn't guessing."

He showed up early. He stayed late when asked.

He learned quickly who to listen to and who to avoid. The discipline drilled into him overseas transferred effortlessly. He followed protocol, handled pressure, and did what was required without complaint. For a while, that seemed to be enough. Then, promotions began to come up.

Lewis said he did not rush toward them at first. He watched how things worked. He paid attention to who advanced and how. He studied the testing process, the eligibility lists, and the rules posted on bulletin boards and tucked away in policy binders.

“That’s when I realized there was a system,” he said. “And then, I realized I wasn’t really in it.”

He took the exams. Passed them. His scores were solid, sometimes higher than those who moved ahead of him. When lists were posted, his name appeared where it should have mattered. Veteran preference was written into policy, spelled out clearly, not as a favor, but as an obligation.

“It wasn’t something they were supposed to decide on,” he said. “It was already decided.”

And yet, when promotions were announced, his name was not called. At first, he assumed there was an explanation. Timing. Seniority. Something he had missed. He waited, the way he had been trained to wait. Quietly. Patiently. Without making noise.

“I thought if I just kept doing my job, it would

correct itself. It did not."

Another round came, another list, and another announcement. Again, Lewis's name stayed where it was while others moved forward. Some had less time in. Some had lower scores. Some had not served.

Lewis noticed something else, too. Conversations stopped when he entered a room. Decisions seemed to happen before meetings were called. He learned things after they were already settled.

"You could feel it," he said. "Like you were always a step behind, even when you were standing right there."

He asked questions carefully. Not accusations. Just questions. He was told to be patient. Told his time would come. Told not to rush the process.

"They always said it real calm," he said. "Like they were doing me a favor."

What unsettled him most was not the denial itself, but the lack of explanation. No one told him he had done anything wrong. No one pointed to a failure or a gap. There was nothing to correct, nothing to improve.

Just waiting.

Lewis said it reminded him of something he

could not quite place at first. The feeling of standing in line, knowing the rules applied differently depending on who you were, of watching others move ahead while being told to stay put.

"It felt familiar," he said. "Too familiar."

Still, he kept working. He kept records. He saved notices. He folded letters and tucked them away. Years in the military had taught him that documentation mattered, even when no one yet acknowledged it.

"I didn't know why I was holding onto it all," he said. "I just knew not to throw it away."

By then, the sleep problems had not gone away. If anything, they were getting worse. He slept lightly, woke often, his body alert in ways the job did not require. Loud noises still caught him off guard. He stayed aware of exits. He scanned the rooms without thinking.

At work, he was steady. At home, his mind stayed busy.

"I didn't have a name for it," he said. "I just knew something wasn't right."

And as the department continued to benefit from his service while withholding advancement, Lewis began to understand that this fight would not be loud or fast.

It would be slow.

It would be procedural.

And it would demand a different kind of endurance than the one that had carried him through war.

Lewis said the first time he heard the words spoken aloud, they did not come from a supervisor or a command officer. It came from someone who leaned in close enough to make sure no one else could hear.

“They know,” the man told him quietly. “They just aren’t going to do anything about it.”

Lewis did not respond right away. He had learned a long time ago that silence could be safer than reaction. But the sentence stayed with him. It settled into place beside everything he had already been noticing.

He asked again, this time formally. Not emotionally. Not angrily. He referenced policy. Veteran preference. Test scores. Dates. He did exactly what the system said a man was supposed to do when something did not add up.

The response was polite, vague, and noncommittal.

They thanked him for his service. Told him the department valued his contribution. Explained that promotions were complex and depended on many factors. None of those factors was named.

"It sounded real good," he said. "But it didn't say anything."

Another cycle passed. Another list went up. Another person moved ahead of him.

Lewis said that was when the waiting changed. It was no longer passive. It became deliberate. He started paying closer attention to how decisions traveled. Who signed what. Who spoke in meetings and who did not. He noticed patterns that had nothing to do with performance and everything to do with comfort.

"They promoted people they were comfortable with," he said. "Not people who were ready."

What stung was not ambition. Lewis was not chasing rank for ego. He wanted stability. A future. A sense that the years he had already given meant something tangible.

"I had already done the dangerous part," he said. "I just wanted the fair part."

Instead, he felt himself being managed rather than led. Encouraged just enough to stay. Discouraged just enough not to push too hard.

At home, the strain showed up in small ways. He slept in pieces. He replayed conversations after they ended. He woke before dawn without knowing why. Some mornings, his body felt like it had already worked a full shift before the day even started.

Still, he showed up.

That consistency became part of the department's quiet expectation. Lewis would be there. Lewis would handle it. Lewis would not make trouble.

"They counted on that," he said. "They knew I wasn't reckless."

But the more dependable he became, the more invisible his advancement seemed to be.

One afternoon, after another announcement passed him by, Lewis sat alone in his car longer than usual before driving home. He did not turn on the engine right away. He sat there, hands resting on the steering wheel, feeling something tighten that had nothing to do with anger. It was clarity.

"I realized then," he said, "this wasn't about time anymore."

He understood something fundamental in that moment. The system was not malfunctioning. It was operating exactly as designed. His patience was not being tested. It was being used. That realization did not make him loud. It made him precise.

Lewis began organizing what he had saved. Notices. Lists. Letters. Dates. Names. He did not yet know what he would do with them, only that they mattered now in a way they had not before.

“I stopped asking when it was going to change,” he said. “I started asking why it hadn’t.”

And that question, once formed, did not go away. It followed him through the hallways. It followed him home. It followed him into sleep that never fully settled. The war had taught him how to recognize a long fight before it announced itself. This one did not come with gunfire. It came with paperwork.

And Lewis understood, with the same certainty that had carried him through jungle and storm, that this would not be resolved by waiting quietly anymore. Lewis said the decision to file a formal complaint was not driven by anger. It came from clarity.

“I kept checking myself,” he told me. “I wanted to be sure it wasn’t just how it felt. I wanted to know what the rules actually said.”

So he did what Marines do. He went through the paperwork. He pulled the postings. The eligibility lists. The exam results. The policies outlining veteran preference. He laid them side by side, the way he once laid maps out before a patrol, looking for what didn't line up.

"It was right there," he said. "In black and white."

Veteran preference was not discretionary. It was not optional. It was not something that could be applied when convenient and ignored when uncomfortable. It was policy. And by that policy, Lewis should have moved forward. Instead, he had been passed over, more than once. So, he filed a formal complaint. It was written, documented, and anchored firmly in the language of their own rules.

"I didn't accuse anybody of racism," he said. "I didn't raise my voice. I just asked one question. Why wasn't the preference applied?"

The response did not come quickly. It came the way institutional responses often do. Slowly. Carefully. Filtered through offices and titles and quiet conversations behind closed doors. And then, eventually, it came in writing.

"They admitted it," Lewis said. "They said veteran preference should have been applied."

He let that sit for a moment when he told me, as if he still needed to hear the words land fully.

"They acknowledged it," he repeated. "They said it plainly."

The acknowledgment did not come with a

correction. There was no retroactive promotion, no adjustment, no apology, and no urgency to repair what they had already admitted was wrong.

"They said it like it was already over," he said. "Like naming it was enough."

Lewis understood immediately what that meant. They were willing to admit error on paper. They were not willing to undo it in practice. And that distinction changed how he understood everything that followed.

Lewis had already been told, quietly and officially, that the system had failed him and had no intention of correcting itself unless forced to.

"That's when I knew," he said. "This wasn't a mistake. This was how it worked."

The complaint did not make him popular. It did not make him protected. But it did something else. It put his name on record as someone who noticed. Someone who read policy closely. Someone who did not accept silence as a resolution. He is a man who knows the system can admit wrongdoing without ever intending to make it right. And that knowledge will shape every fight that follows.

Chapter 7: The Scandal and Internal Fallout

"You can tell when something isn't stable," he said. "Same way you can tell in the field. You don't know exactly when it's going to break. You just know it will."— Lewis Cain.

Lewis said he did not feel vindicated when the department finally admitted that veteran preference should have been applied.

There was no celebration. No relief.

"It didn't change anything," he told me. "They said it was wrong, and then they just kept going."

The acknowledgment came in writing, careful and restrained, the kind of language that sounds official without ever sounding accountable. It confirmed what he had already proven, but it did not move him forward. No promotion. No correction. No attempt to repair what had been denied.

By then, Lewis understood something that did not need to be explained to him twice. Admissions could live comfortably on paper. Decisions lived somewhere else entirely.

"It wasn't a mistake," he said. "It was how they

operated."

He kept reporting to work. He kept his uniform pressed. He did the job the same way he always had. But something inside the department had shifted. He said he could feel it before he could point to it.

"That's when things started to unravel."

It did not begin loudly. There was no announcement, no crisis meeting, no public fracture. It began in smaller ways. A promotion that moved unusually fast. A name was placed on a list that did not quite align with policy. A decision that felt rushed, rather than earned.

Lewis noticed tone before he noticed detail. Doors closed more often. Conversations ended when certain people entered a room. Leadership, which had once hidden behind the complexity of rules, suddenly moved with surprising ease when it suited them.

"There were people being pushed forward who weren't ready," he said. "Everybody saw it."

One of those names was Renee Dawson.

When he said her name, there was no edge in his voice. No accusation. Just a fact. She had moved quickly through the ranks, faster than others with more years and stronger test scores. The explanations were brief. The justifications thinner still.

“She was protected,” he said. “That part was obvious.”

At first, it looked like favoritism. The kind departments defend as vision or confidence in emerging leadership. But Lewis had been inside systems long enough to know that when rules bend for one person, they do not stay bent. Eventually, something gives.

“You can tell when something isn’t stable,” he said. “Same way you can tell in the field. You don’t know exactly when it’s going to break. You just know it will.”

What had once been quiet and procedural was about to become public. And Lewis, whether anyone acknowledged it or not, had already been standing at the fault line. Lewis said he was at home when the call came.

“I never understood how they justified it,” Lewis told me. “They pushed Renee through like the rules didn’t apply.”

He had clapped for her at her ceremony. He had meant it. He wanted her to succeed. He believed that when one of them advanced, it meant doors were opening. But the speed of her advancement, the insulation around her decisions, and the way concerns were brushed aside had unsettled him long before that night.

The phone rang later than routine calls usually did. He answered, expecting something administrative. Instead, he heard urgency in the voice on the other end, the kind of strain that flattens formality.

“There’s been a shooting.”

For a moment, the words did not fit. The jail was supposed to be a controlled space. Weapons did not cross those thresholds. Inmates did not gain access to firearms inside secured corridors.

“What do you mean by a shooting?” he asked.

An inmate had a gun.

Lewis said that was the part that altered everything.

“A gun doesn’t just appear inside a jail,” he said quietly.

He could hear chaos in the background of the call. Radios overlapping. Boots striking concrete. Commands delivered too fast to process. Deputy Michael Delaney had been shot in the back of the neck. The building had gone from routine to emergency in seconds.

When the call ended, Lewis remained standing in his living room, the receiver still in his hand. He did not sit down right away. He did not speak. He understood immediately that the real question would not be about the shot. It would be about the weapon.

By the time the first reports began circulating, the answer was already spreading ahead of official confirmation. It moved through the department in low voices before reaching the media.

Renee Dawson.

Lewis did not react with anger. He reacted with a heaviness he could not easily name.

"I felt bad for her," he said. "I really did."

He could not reconcile how she had allowed it to happen. He could not understand the decision. But, he also could not ignore the fact that the same system that had stalled him had accelerated her without hesitation. They had endorsed her and protected her. Positioned her. And now a man lay wounded, and another deputy was facing criminal charges.

He turned on the television when the breaking news banner cut across the screen.

The anchor's voice carried that urgent, polished tone reserved for scandal. The words scrolled beneath her image in bold letters. A firearm had been smuggled into the county jail. An inmate had used it to shoot a deputy. An internal investigation was underway. Deputy Renee Dawson's name was being linked to the weapon.

Lewis watched her photograph appear on the screen. The same uniform. The same smile from her ceremony. Now framed by language that felt final.

"They exposed it," he said. "But it came at a cost."

He was not glad about the way it happened. He was not satisfied. There was no sense of vindication in seeing the department publicly shaken. A deputy had been shot. Another deputy was going to jail. Whatever confirmation this provided about the instability he had sensed, it was stained by blood.

"They should have fixed it when it was still quiet," he said. "It didn't have to happen like that."

Inside the department, the reaction was immediate and strained. The halls no longer carried routine conversation. They carried tension. Deputies who once moved easily now spoke in fragments. Leadership began moving quickly, not to acknowledge what had led to the failure, but to contain it. Files were pulled. Statements drafted. Narratives shaped.

Lewis walked through that building in the days that followed and felt something different in the air. The confidence that had once accompanied

Dawson's rise had evaporated. The certainty that had shielded her decisions was gone. In its place was a quiet scrambling.

"They pushed her through," he said. "And then they acted surprised when it collapsed."

He did not celebrate being right. He did not take satisfaction in the exposure. What he felt was sorrow layered with clarity. He had warned in policy language. He had documented inconsistencies. He had pointed out that veteran preference was being ignored. The system had acknowledged wrongdoing and moved forward anyway. Now the same system stood publicly fractured.

"This wasn't just about my promotion anymore," he said. "This was about what happens when you ignore the right person and protect the wrong process."

He had once believed the fight was procedural. That night, he understood it was structural. And structural failures do not break quietly.

In the days after the shooting, the department moved as if it were trying to hold its own breath.

Lewis said the halls felt narrower, though the walls had not moved. Conversations that once rolled easily through open doorways now collapsed into whispers. Men who had debated policy openly now leaned in close, speaking low enough that even the air seemed excluded. It was not just tension; it was calculation.

"They weren't talking about what happened," he told me. "They were talking about how to survive it."

Official statements were drafted quickly. The language was controlled, precise, and careful to protect the institution. Investigations were announced. Internal reviews promised. But Lewis noticed something else. The same leadership that had insisted promotions were complex and nuanced when he questioned them now moved with urgency when public scrutiny arrived.

Inside the department, Dawson's name became both unavoidable and unspeakable. Some deputies expressed anger openly. Others avoided the topic altogether. A few looked at Lewis differently, not with apology, but with a kind of restrained awareness.

"They knew," he said. "They just weren't going to say it."

No one approached him to acknowledge that he had raised concerns about policy long before the scandal. No one connected the pattern aloud. Instead, the distance between him and leadership widened. Meetings occurred without him. Information traveled around him. Assignments shifted quietly.

"It got colder," he said. "Not loud. Just colder."

He began noticing small exclusions. Conversations that stopped when he entered a room. Invitations that never came. Supervisors who once offered routine updates now kept exchanges brief and strictly functional. The message was not explicit, but it was clear enough.

He had become inconvenient.

"They didn't want someone in the room who remembered what the rules said," he told me.

The complaint he had filed months earlier now lived differently in the department's memory. Before the scandal, it had been manageable, procedural, something that could be acknowledged and contained. After the shooting, it became something more dangerous. It suggested that instability had been visible before it became violent.

And Lewis was the one who had documented it.

No one accused him of anything. No one confronted him directly. But the shift was unmistakable. Supervisors' eyes lingered longer. Evaluations became more meticulous. Minor infractions were suddenly scrutinized with unusual precision.

"They started watching me," he said. "Not because I did something wrong. Because I noticed."

He did not respond with anger. He responded with discipline. He arrived earlier. He documented more carefully. He saved copies of everything. Years in the Corps had taught him that when attention tightens around you, your posture must steady.

At home, he reflected on what had unfolded. He felt genuine sorrow for Dawson. He thought about the ceremony where he had clapped for her, about the pride he had felt seeing someone who looked like him move forward. He could not reconcile the decision she had made. Still, he also could not ignore the environment that had elevated her without correction or challenge.

"They set her up," he said quietly. "And then they let her fall."

He did not excuse her actions. A deputy had been shot. A weapon had entered a secured facility. Those were facts that carried their own significance. But Lewis understood something else. Systems that ignore structure for appearance eventually expose both.

"They thought I would feel vindicated," he said. "I didn't. I felt confirmed."

That confirmation did not make his days easier. It made them heavier. He walked into work aware that the same system that had denied him quietly now regarded him as something more complicated.

He said the night Deputy Delaney was shot did not end when the news cycle moved on. It followed him home. It followed him into sleep.

"I kept seeing Reed," he told me.

Not in some abstract way. Not as a memory he chose to revisit. Reed returned uninvited. The rain. The sound of gunfire cutting through the jungle. The way Reed had clutched his throat when he went down.

"When I heard Delaney got hit in the neck," Lewis said, "it took me right back."

He did not say it dramatically. He said it as fact. The body does not separate decades the way the mind tries to. A bullet is a bullet. A throat is a throat. The sound of a man going down does not age out.

He began waking up more frequently after that night. Not just restless, but alert. Sitting upright before he fully understood why. His heart was already racing. The room was unfamiliar for a split second before it came back into focus.

"I wasn't dreaming new things," he said. "I was reliving old ones."

At work, the atmosphere grew colder. The scandal had exposed the department publicly, but internally, it hardened something. Conversations shifted. The language changed. Jokes that once hid behind subtlety sharpened.

Lewis began noticing comments that did not need explanation.

"You people are always making things complicated."

"Why can't you just let things go?"

"You already got your little complaint acknowledged."

He said the word little stuck with him.

Promotions continued to stall. Lists came and went. Lewis's name appeared and remained. Others advanced. The explanations grew thinner, and the patience they expected from him grew thicker.

"It didn't matter what my file said," he told me. "It mattered what I looked like."

He witnessed decisions that cut corners on constitutional protections. Shortcuts justified as efficiency. Stops that did not quite align with probable cause. Arrests that felt more personal than procedural. When he raised questions, even carefully, the temperature shifted.

"They started calling me difficult," he said. "Not wrong. Just difficult."

The word followed him into evaluations. Into

informal conversations. Into assignments that placed him slightly outside the center of things. He was no longer simply the deputy who knew policy. He was the deputy who questioned. And questioning, in a department recovering from scandal, was unwelcome.

"They wanted loyalty," he said. "Not memory."

He kept working.

He kept documenting.

He kept watching.

But something else was happening beneath the surface.

His body began responding before he consciously registered why. Loud noises inside the jail made him flinch harder than before. A slammed door echoed longer than it should have. His eyes scanned corridors automatically, calculating distance to exits, assessing angles, and noting who stood where.

"I was on edge all the time," he said. "Like something was about to go wrong."

The scandal had confirmed instability. The racism had confirmed exclusion. The stalled promotions had confirmed deliberate obstruction. His mind tried to process it all through discipline. His body processed it through adrenaline.

Sleep shortened. Nightmares intensified. Reed appeared more frequently now, not just as a memory, but as a presence. Sometimes he would see him at the edge of his vision, standing still, silent. Once, he said he woke up convinced he heard boots running down his hallway, only to find the house empty.

"I didn't have a word for it," he told me. "I just thought I wasn't sleeping right."

He became more guarded. Quicker to react. Slower to trust. At home, he found himself sitting where he could see the door. In restaurants, he preferred corners. He listened to conversations around him even when he did not mean to.

The department saw an increasingly serious deputy. What they did not see was a man whose nervous system had never returned from Vietnam and was now being triggered by institutional betrayal.

"The war didn't end," he said quietly. "It just changed locations."

Inside the department, he was isolated.

Inside his body, he was mobilized.

And both battles were escalating simultaneously.

Lewis said the confrontation did not begin

loudly.

It started in a briefing room after shift change, the kind of room where deputies leaned back in metal chairs and pretended not to be listening too closely to one another. The shooting had been weeks earlier, but its shadow still lingered. Promotions were still being discussed in hushed tones. Assignments were being reshuffled.

Lewis had raised a procedural question about an arrest from the previous evening. Not accusatory. Just precise. He referenced policy, the way he always did.

The room tightened.

A senior deputy leaned forward, elbows on his knees, and stared at Lewis longer than necessary.

“You always got something to say, don’t you?”

Lewis did not respond immediately. He kept his voice even.

“I’m just asking if we followed the guidelines.”

The deputy gave a short laugh that did not carry humor.

“Guidelines. Policy. Preference. You sound like you think you’re above everybody else.”

Lewis felt the shift before the words fully landed.

"I'm not above anybody," he said. "I just want it done right."

That was when it changed.

The deputy's voice dropped lower, but harder.

"Maybe if you people stopped making everything about rules and race, things would run smoother around here."

The room went still.

Lewis said he did not remember standing up. He only remembered that suddenly he was on his feet, his body reacting before his mind finished processing what had been said.

"You people."

The phrase did not feel new. It felt inherited. It felt like Mississippi. Like men who spoke calmly while holding power. Like dogs straining at leashes and officers standing ready behind them.

"I'm talking about procedure," Lewis said, his voice controlled, but tight. "You're the one talking about race."

The deputy stood now, too.

"What I'm talking about is you stirring things up every time you don't get your way."

There it was, the reduction he had come to

recognize over the years, the subtle reframing that transformed a documented policy complaint into something personal, something driven by ego rather than principle, as if his insistence on written rules had never been about procedure at all but about pride. Lewis said his heart was pounding so hard he could feel it in his throat. For a split second, the room blurred. Not metaphorically. Literally. The walls seemed to close in at the edges. The fluorescent lights felt too bright.

His body reacted before his mind did. He felt his jaw tighten. His hands curled inward. His shoulders shifted slightly, instinctively, the way they had in Morocco when a white man from Alabama had stood over him with a bottle and said he was going to "whoop all you boys."

"I felt that same energy rise up," Lewis told me. "The same surge."

He did not see Reed this time.

He saw the bottle coming towards his head.

He saw the quick movement of his own hand reaching for the switchblade.

"That same pressure hit me in that room," he said. "Like my body was already moving."

The instinct was immediate and physical.

Neutralize the threat.

That is what his body remembered.

The deputy shouted.

"You're always stirring something up, Cain. Maybe that's why you're still where you are."

He could feel the adrenaline moving through him, fast and electric. His muscles were ready. Balanced. Alert.

But this was not Morocco.

This was Wisconsin.

He forced his hands open. He slowly forced air into his lungs.

Someone between them muttered, "That's enough."

The deputy shook his head, but stepped back.

The meeting dissolved into discomfort as chairs scraped against the floor. Papers were gathered with unnecessary focus, men deliberately avoiding eye contact as if silence could smooth over what had just occurred, and although nothing had been resolved, the confrontation was quietly absorbed into the room rather than addressed. Lewis walked out of that room with his pulse still elevated. By the time he reached his car, his hands were trembling slightly. He stared at the steering wheel for several seconds before starting the engine.

"I wasn't scared of him," he told me. "That wasn't it."

It was the recognition of how close his body had come to reacting the way it once had. The same surge. The same lethal clarity. The same split-second instinct to respond with force.

"I knew if I didn't get control of it," he said quietly, "something was going to happen."

Lewis told me that night he did not sleep. His mind replayed the moment in the briefing room and the moment in Morocco side by side. The bottle. The blade. The words "you people".

The department saw a deputy who had stood his ground. But Lewis felt like a man who had barely contained himself. The next morning, without telling anyone at work, he drove to the VA Medical Center.

"I couldn't ignore it anymore," he said. "My body was still fighting wars."

Lewis said walking into the Lakeside VA Medical Center felt different from walking into a jail or a courtroom.

There were no ranks to measure. No supervisors to outmaneuver. Just a waiting room full of men sitting quietly with their own histories folded into their posture. Some stared at the television mounted high in the corner. Others stared at nothing.

"I didn't think I belonged there," he told me. "I thought I just needed sleep."

He checked in, filled out the forms, and wrote "Vietnam" under service history without elaboration. When the nurse called his name, he followed her down a hallway that smelled faintly of disinfectant and old paper files. The walls were neutral. Calm. Designed not to provoke anything.

But his body stayed alert.

When he sat across from the doctor for intake, he kept his back straight and his hands resting flat on his thighs. The physician, a middle-aged man with wire-framed glasses, scanned the chart briefly before looking up.

"What brings you in today, Mr. Cain?"

Lewis hesitated.

"I haven't been sleeping right," he said.

The doctor nodded as if that were common.

"How long?"

"Years," Lewis replied. Then he corrected himself. "Worse lately."

The doctor asked about his appetite, his work,

and his alcohol use. Lewis answered plainly, offering no embellishment and no dramatics, only the facts as they were.

"Any nightmares?"

Lewis shifted slightly in his chair.

"Sometimes."

"About Vietnam?"

"Yes."

The doctor made a note.

"Do you ever feel on edge? Hyperalert?"

Lewis almost laughed at the word hyperalert.

"I don't turn it off," he said. "I don't know how."

The physician leaned back slightly, studying him now rather than the paperwork.

"Have you ever felt like you were back there? Even briefly?"

Lewis did not answer immediately. He thought about the briefing room. About the phrase you people. About the rush of adrenaline that had felt indistinguishable from Morocco.

"Yeah," he said finally. "I have."

The doctor asked him to describe one instance.

Lewis told him about the confrontation at work,

not the policy dispute or the politics behind it, but the surge of adrenaline, the narrowing of his vision, and the physical readiness that took over before his mind had time to process what was happening.

"I felt like I was somewhere else for a second," Lewis said. "Like I had to react."

The doctor nodded slowly.

"That's not uncommon," he said carefully. "Especially in combat veterans."

He asked about irritability, difficulty concentrating, and sudden anger, and Lewis acknowledged that he had been shorter with people lately, more tense and less patient than he used to be.

"What about intrusive memories?" the doctor asked. Lewis thought about Reed. About Morocco. About the bees. About the bottle.

"They come when they want to," he said.

The doctor briefly closed the file and folded his hands.

"Mr. Cain, what you're describing is consistent with post-traumatic stress."

Lewis had heard the term before, loosely, in passing, delayed combat fatigue, and battle stress. Words that sounded temporary, yet manageable.

"I'm not crazy," Lewis said flatly.

"I didn't say you were," the doctor replied evenly. "This is a neurological response to trauma. Your body learned to survive extreme threat. It hasn't fully unlearned it."

Lewis absorbed that without reacting.

"So, what does that mean?"

"It means your nervous system is still operating as if the threat is present. Loud noises, confrontations, stress, perceived injustice; they can trigger the same physiological response you experienced in combat."

The doctor continued.

"It doesn't make you weak. It means you adapted to survive. The problem is, the adaptation doesn't switch off automatically."

Lewis stared at a spot on the wall just behind the doctor's shoulder.

"I've been working this whole time," he said. "Holding a job. Doing it right."

"And that matters," the doctor said. "Functioning doesn't cancel out trauma."

The word trauma felt heavy.

The doctor recommended further evaluation,

counseling, and possibly medication to help with sleep. He explained that symptoms often intensify under stress, particularly when a person feels threatened or invalidated. Lewis almost nodded at that.

"It's been stressful at work," he admitted.

The doctor did not press for details. He did not need to.

"I'd like to schedule you for a full psychiatric evaluation," he said. "We'll document everything."

Lewis understood documentation. He left the appointment with a follow-up date, a prescription for sleep assistance, and a new word attached to his experience—post-traumatic stress.

"I didn't like it," he told me. "But I knew it fit."

Lewis said that when he stepped outside the VA, he did not get into his car right away. He stood beside it, the metal warm under his hand, and drew in the cool air slowly, filling his lungs as if he needed proof that he could. The parking lot was ordinary, cars idling, doors opening and closing, but inside his chest, everything was moving too fast.

Memories rose without order. Vietnam. The elephant grass pressing in around him, thick and unforgiving. The confrontation in the briefing room. The phrase "you people". Morocco. The bottle. The blade. Each moment, competing for space in his mind, none of them willing to yield.

"I could feel my face starting to tingle," he told me. "Like the blood was rushing up into it."

His heart pounded hard enough that he was certain someone standing close would hear it. For a moment, the air felt thin again, the way it had in the jungle when he pushed through tall grass that closed in from every side, the heat wrapping around his body, the sense that the space ahead was shrinking instead of opening.

"It felt like I couldn't get a full breath," he said.

He leaned against the car and forced himself to inhale deeply, steadying his body the way he had learned to do years ago under far different circumstances. The building behind him was calm, clinical, and efficient. Inside it, his symptoms had been categorized, documented, and named.

Post-Traumatic Stress Disorder.

He repeated the words to himself, not as defeat, but as recognition.

"The war had followed me home," he said. "I just finally had a word for it."

Chapter 8: The Percentage of a Man

"That's when I started paying attention to everything," he said. "Because if they could put it in writing and still act like it didn't matter, then the fight wasn't about diagnosis. It was about what they were willing to honor."— Lewis Cain.

"So when did it become official?"

He leaned back when I asked it, not defensive, not hesitant, just searching for the right doorway into the memory. The room was quiet except for the faint hum of the air vent. For a moment, it felt as if we were both sitting in that VA office, decades earlier, waiting for someone else to decide what his mind and body were worth.

"It became official," he said slowly, "the day they wrote it down."

The full psychiatric evaluation took place at the VA in Lakeside. The waiting room held men who spoke little, men who stared at televisions mounted in the corners, but did not seem to see what was playing. Lewis had already been seen before, already told someone about the nightmares, the sleeplessness, the tension that never fully left his shoulders. This time was different. This time was comprehensive. Structured. Measured.

The psychologist sat across from him with a yellow legal pad and a chart thick enough to suggest history. The questions were methodical. Childhood. Military service. Combat exposure. Employment history. Sleep patterns. Irritability. Concentration. The doctor's tone was neutral, neither skeptical nor sympathetic, simply gathering.

Then, the questions shifted.

"Do you ever see or hear things other people do not?"

Lewis told me he did not answer right away. Not because he did not understand the question, but because he understood exactly what it meant. Words like that could follow a man for the rest of his life. Words like that could be used to explain him away.

He chose not to protect himself.

"Yes," he said.

Lewis went on to explain in detail. Sometimes he saw Reed. Not as a ghost, not as something floating in the room, but as an overlay. A sudden flash of Reed's face in the place of another man's. A body on the ground where there was only carpet. A split-second merging of present and past that his mind corrected almost immediately, but not before his heart reacted.

Sometimes he heard the sounds of Vietnam without warning. The snap of gunfire in the distance when a car backfired. The rush of movement in elephant grass when the wind pushed through the trees. Sometimes his body moved before his thoughts caught up, muscles tightening, breath shortening, eyes scanning exits before he consciously decided to look.

"I knew where I was," he said. "I wasn't confused about that. But it would hit me, and for a second I was there again."

The psychologist did not interrupt. He did not overreact. He wrote.

Combat exposure. Recurrent intrusive imagery. Hypervigilance. Startle response. Sleep disturbance. Hallucinatory overlays tied to trauma memory.

The questions continued, but the tone had shifted; this was no longer about whether Lewis was stressed or struggling to adjust. It was about permanence. About impairment. About what the war had carved into him and refused to release.

When the evaluation ended, the psychologist closed the file and told him the findings would be forwarded for formal determination. Weeks later, the letter came. Clinical language. Structured paragraphs. No embellishment.

Combat-related Post-Traumatic Stress Disorder. Chronic. Severe. Permanent impairment.

Lewis held the letter in his hands and read it several times. It did not feel like validation. It felt like a translation. For the first time, what had been living inside his body had been reduced to terminology the system understood.

"That was the day," he said. "That's when it stopped being something I was dealing with and started being something they acknowledged."

That evaluation became the foundation for everything that followed. It moved from a doctor's desk to an official record, from observation to declaration. It would later be cited in correspondence and filings, stamped and forwarded through channels that carried legal importance.

The internal war had finally been named in writing. Now the system had to decide what that name was worth.

He did not just tell me about that evaluation. He showed me.

Lewis disappeared briefly into the back room of his house and returned carrying a thick black binder so overstuffed it barely closed. The edges were worn. The rings inside strained against their hinges.

Loose papers protruded at uneven angles, some tucked carefully into plastic sleeves, others folded and refolded from years of handling.

He placed it in front of me on the kitchen table without ceremony.

"Start there," he said.

The binder was heavier than it looked. Not because of the paper, but because of what it represented. I opened it slowly, half expecting chaos. What I found instead was structure. Tabs divided sections. Medical in one grouping. Legal in another. Correspondence in its own pocket. Others had handwritten notes in the margins in a steady, controlled script. Dates were underlined. Paragraphs were highlighted, and key phrases circled.

Combat-related.
Chronic.
Permanent.

There were VA appointment summaries. Psychological assessments. Letters stamped and initialed. Copies of evaluations that had been forwarded to state agencies. Envelope fronts preserved, with postmarks visible, as if Lewis had long ago understood that timing would matter.

"It's all in there," he said quietly. "Everything they said. Everything they filed. You can take it home. I have duplicates of everything."

As I turned the pages, I felt the credence of repetition. Visit after visit. Documentation layered on documentation. The language clinical and consistent. PTSD. Severe. Impairment. Occupational limitation. Recommended disability consideration.

It was overwhelming to take it all in at once. Not because it was disorganized, but because it was thorough. The binder was not the work of a man grasping at memory. It was the work of a Marine who understood chain of custody, who knew that if something mattered, it had to be recorded. This was not an obsession. It was defense.

"I kept everything," Lewis said. "I knew I'd need it."

That binder was the physical embodiment of what the full psychiatric evaluation had started. The war had been named in ink. The impairment had been documented. The system had been informed.

And Lewis had made sure the record could never quietly disappear.

Lewis turned several pages in the binder before stopping. The plastic sleeve crackled as he carefully slid the document out, handling it as if it might tear under too much pressure. The paper had yellowed slightly at the edges, but the print remained clear.

Dated May 20, 1991.

Signed by a County-appointed evaluating physician.

The language was direct and clinical, with no exaggeration or ambiguity—permanent disability.

The physician concluded that Lewis's psychiatric impairment was chronic, severe, and directly connected to his military service. It stated that his condition significantly limited his occupational functioning and would continue to do so indefinitely.

Two days later, the evaluation did not remain in a file drawer. A letter dated May 22, 1991, authored by a senior County legal representative, forwarded the physician's findings to the State Department responsible for labor and industry matters. The transmission made the determination part of the official state record in Lewis's compensation case.

"They knew," he said. "From that point on, they knew."

The forwarding letter clearly referenced the medical conclusion. It acknowledged the evaluation. It requested that the report be included in the active file. The paper trail was clean. Sequential. Unbroken.

"This should have settled it," he said. "Once

they call you permanently disabled, that's not a small thing."

Permanent disability was not symbolic. It carried financial implications. Legal implications. Dignity implications. It meant a veteran who could no longer function at full capacity because of service-connected trauma would be supported accordingly. But after the evaluation was filed, something unusual happened.

Nothing.

No formal notice implementing permanent benefits.
No correspondence explaining next steps.
No adjustment in compensation consistent with the determination.

The record reflected acknowledgment.

The outcome did not reflect correction.

Lewis closed the binder slowly.

"That's when I started paying attention to everything," he said. "Because if they could put it in writing and still act like it didn't matter, then the fight wasn't about diagnosis. It was about what they were willing to honor."

Several months after the permanent disability evaluation had been forwarded to the state, Lewis told me he discovered another letter in his file.

According to Lewis, it was dated September 11, 1991, and had been authored by his attorney at the time. Lewis said that when he read it, the letter appeared to represent that he had agreed to a compromise settlement with the County.

Lewis told me he had never seen or received any such agreement.

He said he had never signed one, never approved one, and had never been presented with any terms he agreed to accept. He did not recall the matter being put before him at all.

Yet, Lewis said, the letter existed in his file. And according to his account, it would later be referenced in proceedings as though it represented a concluded understanding between the parties — one he maintained he had never reached.

When Lewis first discovered it, he did not react with anger. He reacted with disbelief. He returned to his binder and began cross-checking dates. The May 20 permanent disability determination. The May 22 forwarding letter. The absence of implementation. Then, the September correspondence implying a settlement.

No signed settlement agreement was attached.
No transcript reflecting his consent.
No recorded acknowledgment bearing his signature.

Just the attorney's written representation.

"It read as if I'd agreed to something," Lewis said. "But, I hadn't."

The permanent disability evaluation was no longer treated as an unresolved obligation but as a matter implicitly settled. The language shifted subtly. Instead of addressing whether the County had honored the physician's determination, proceedings began to reference the supposed compromise as though it explained the absence of further benefits.

Lewis kept searching the file for the document that would make it make sense.

There was none.

"It felt like I was being rewritten," he said. "Like the file said something different than what I lived."

That was the moment the binder changed from archive to armor.

Lewis began organizing everything with greater precision. Medical records in one section. Legal correspondence in another. Hearing notices, court orders, and envelopes preserved with their postmarks intact. He underlined dates not to dramatize them, but to preserve sequence. If someone claimed a settlement existed, he wanted to be able to trace the claim back to its source and test it against the timeline.

As Lewis understood it, the system did not deny that the physician had declared him permanently disabled. What he observed, and what his documentation appeared to support, was that the declaration had been absorbed into a framework that treated the matter as already resolved.

Lewis closed the binder when he finished walking me through that portion.

“They didn’t argue the diagnosis,” he said. “They just acted like it didn’t require anything.”

From that point forward, every hearing, every order, every procedural shift would have to be measured against one question: if permanent disability had been declared and filed, why had it not been implemented?

The financial shift did not come with an explanation. It came quietly, through numbers.

For years, Lewis’s monthly pension had reflected what he understood to be his rightful entitlement based on service and disability status. It was not extravagant, but it was steady. It represented, however imperfectly, an acknowledgment that his body and mind had paid a cost.

Then, the amount changed.

What had once been approximately $1,500 a

month was reduced to roughly $400.

No detailed breakdown accompanied the reduction. No formal notice explaining how the calculation had been altered in light of the permanent disability finding. Just a smaller number appearing where a larger one had been.

Lewis told me he initially assumed it was a clerical error.

"I thought maybe something had been misapplied," he said. "Because you don't cut a man's pension by more than half and not explain it."

He called. He wrote. He requested clarification. The answers were procedural and fragmented, often referring him back to prior filings or decisions that did not directly address the discrepancy. What should have been a straightforward accounting became layered in administrative language.

Over time, the reduction was treated as a settled fact.

The permanent disability determination remained in the file. The alleged compromise letter remained referenced. But the pension amount reflected neither clarity nor consistency with the medical conclusion that he was permanently impaired.

"That's when I understood," Lewis said. "This

wasn't just about paperwork. They were wearing me down."

The financial strain did not stay confined to spreadsheets and bank statements. It moved into his body.

At first, the changes were subtle. Lewis told me he began feeling tired in a way that sleep did not correct. Not the restlessness he associated with PTSD, but a heaviness that lingered even on mornings when he had managed a few uninterrupted hours. His ankles would swell by evening, socks leaving deeper impressions against his skin. He noticed that his blood pressure readings were creeping upward during routine appointments, numbers that doctors flagged, but did not yet treat as urgent.

"I thought it was just stress," he said. "And it was. But it was more than that."

The stress was not episodic. It was sustained. Legal correspondence was stacked on his desk. Phone calls that led to other departments. Letters that referenced prior decisions without addressing his questions. Each appeal required documentation, and each response required waiting. The reduction in pension meant he had to be more cautious with every expense, calculating what could be postponed and what could not.

Chronic stress does not announce itself dramatically. It accumulates.

At follow-up appointments, lab panels began showing early irregularities. Elevated creatinine. Borderline kidney function. Doctors advised monitoring. Hydration. Blood pressure control. At that stage, no one connected it to anything larger. It was treated as an emerging issue, something to watch.

Lewis listened carefully during those visits. He asked questions. He wrote things down.

“The numbers were changing,” he said. “I could see it on the page.”

The legal fight continued. The pension remained reduced. The permanent disability determination remained on file without meaningful financial correction. And as the years stretched forward, the strain between recognition and implementation became a steady pressure he carried daily.

The workers’ compensation board did not move quickly. Administrators changed. Attorneys came and went. Hearings were scheduled, rescheduled, and sometimes resolved in ways that generated more paperwork than clarity. Lewis learned procedural language the way he had once learned field commands. Motions. Petitions. Responses.

Interlocutory orders.

While he documented the system, his body was documenting him.

Blood pressure remained elevated despite attempts to control it. Fatigue deepened. There were days when concentration required more effort than it had before. Doctors began asking about family history. Diet. Lifestyle. They monitored kidney function more closely.

No one yet mentioned Camp Lejeune.

At that stage, the medical focus remained immediate and individual. Manage stress. Control blood pressure. Reduce strain. But the stress was not reducible. It was built into the structure of the fight.

"There's a kind of pressure that doesn't let up, it's constant," Lewis said.

The pension reduction was never restored to its previous level. Appeals continued. Clarifications were requested. Responses arrived slowly or incompletely. The permanent disability determination remained part of the official record. Yet, the financial outcome did not align with it in a way that resolved the discrepancy.

His faith became routine, rather than dramatic. Scripture before correspondence. Not to erase the stress, but to keep it from reshaping him into something reactive. He believed the system was counting on exhaustion, attrition, and the possibility that, over time, he would stop pressing for correction.

Instead, he documented more carefully. At the same time, his labs continued to shift. It would take years before anyone would ask the question that tied it all together. For now, the symptoms were treated as an emerging chronic illness under stress.

And Lewis continued to fight on paper, even as something inside him slowly failed.

Chapter 9: The Poison that Traveled with Him

"I thought about all the times I drank that water," he said. "Every day. We didn't question it. We trusted it."—Lewis Cain

Lewis shifted in his chair when he got to this part.

Not dramatically. Not with the broad physical signal that telegraphs emotion before language can carry it. It was smaller than that. His hands, which had been resting open on the table the way a man rests his hands when he is certain of what he is saying, moved inward. His forearms pressed together slightly. His posture tightened by a degree that most people would not have noticed.

I noticed.

By then I had spent enough hours across from Lewis Cain to know that his body spoke before he did. The hands told you what the voice was still organizing.

"It started slow," he said.

He was not talking about a legal case. He was not talking about a system, a document, or a department that had failed him. He was talking about something that had no office address, no paper trail, no official to confront. He was talking about what was happening inside him.

That distinction mattered. I let him take his time.

He was a disciplined man about his body. I had seen the exercise bike in his living room during my first visit, not pushed against a wall, the way unused equipment tends to end up, but positioned in the center of the room like a piece of furniture that had earned its place. He used it. He had always used something. The Corps had built that into him, and it never left. You do not train a man to treat his body as a precision instrument for forty years and then watch that instinct simply stop. Lewis ate carefully. He monitored what he put in. He did not smoke. He rarely drank. His body was a record of intentional care across eight decades, and he knew it the way a mechanic knows an engine he has maintained himself, intimately, specifically, by the sound of what normal feels like.

So when something changed, he knew.

“It was the fatigue first,” he said. “Not tired. Not the kind where you sleep, and it goes away. This exhaustion was something else.”

He searched for the word and then gave up on finding a better one.

“Heavy,” he said. “Everything felt heavy.”

He had climbed stairs his whole life without thinking about them. Now he thought about them. Not because they hurt, but because somewhere between the second and third floor his body seemed to recalculate, seemed to request something from reserves he had always assumed were deep. He would pause on a landing and not understand why. A man who had run in formation through sand before dawn should not need to pause on a landing. He knew that. The knowing made it worse.

His blood pressure had always been something the doctors watched. That was not unusual for a man his age who had carried what he carried. But it had been manageable. Numbers that responded to the adjustments. What had worked before stopped working. He would take his medications, do what he was told, eat the right things, drink the water, and the numbers would come back elevated anyway, as if something inside him had stopped taking the instructions.

"The doctors kept saying monitor it," he told me. "And I kept monitoring it. And it kept going the wrong way."

He said that quietly. No bitterness, no accusation. Just the flat recording of a fact that refused to cooperate with him no matter how carefully he responded to it.

He tried to assign it a cause. That is what a logical man does: he finds the origin, addresses it, moves forward. The most logical candidate was the accumulated pressure of Wisconsin. The discrimination had not been occasional. It had been structural, built into the institution he reported to every day, embedded in promotions that moved around him, in the language of certain briefing rooms, in the kind of silence that descends when a particular man walks in. Carrying that kind of awareness for years is a physical act, not just a psychological one. The body grades everything. It registers what the mind tries to file away.

And on top of that, the pension fight, the workers' compensation hearings, the legal correspondence stacking up on a desk that never cleared. Decades of fighting systems that pretended not to see what was plainly visible.

It was reasonable to conclude that his body was presenting the bill for all of it.

But it did not behave like a bill that could be paid.

He decided to leave Wisconsin the way a man decides to step back from a fire. Not in panic, but with the clear-eyed recognition that proximity to this particular thing is costing him something he cannot get back.

"I thought if I could get distance from all of it," he told me, "my body would have a chance."

He said "a chance" the way people say it when they have already stopped believing in certainty.

He chose Savannah.

The warmer climate was part of it. Coastal air. A slower pace. A geography that did not carry Wisconsin in every familiar street corner and every office building he had spent years fighting inside. He wanted somewhere his nervous system could stop bracing.

When he arrived, the city was beautiful in the way Savannah is beautiful; live oaks draped in Spanish moss, the squares, the unhurried quality of the light in the late afternoon. He let himself settle into it for a moment.

"It was quieter," he said. "At least at first."

But the South has its own grammar, and Lewis had been raised in Mississippi. He knew how to read it. The conversations that shortened when he entered certain spaces. The eyes that assessed before they engaged. The unspoken boundaries that required no signs because everyone who grew up inside this particular history already knew where they were. He had not left that knowledge behind in Jackson. He had carried it his whole life, and it told him now, plainly, that Savannah was not the distance he had been looking for.

He was not afraid of it. But he understood, after everything his body had already been through, that he did not have the surplus to spend on it.

"I thought I was getting away from it," he said. "But I wasn't."

And underneath all of that, the more pressing fact, his body was not responding. The blood pressure remained elevated in Savannah the same way it had in Wisconsin. The fatigue did not ease in the coastal air. The kidney numbers, which doctors had been describing as something to watch, were moving further in the wrong direction regardless of where he was. The distance from

Wisconsin changed the scenery without altering the trajectory.

"It didn't get better," he told me. "That's when I knew it wasn't just where I was."

He paused when he said that. I watched him work through the implications of it, the way you watch someone do math in their head and arrive at a number they weren't expecting.

He left Savannah.

Nashville came next. Mt. Juliet, just outside the city, quiet, set back from the pace of downtown. This place offered the practical access he needed and something closer to stillness. He was not chasing hope anymore by that point. He was making calculated decisions about what his body required.

"I wasn't trying to be in the middle of everything," he said. "I wanted somewhere quieter."

For a time, it was as he settled in. He kept his appointments. He maintained his routines. He continued taking his medications the way a man takes them when he still believes compliance will eventually produce results.

The results continued to move in one direction.

His kidney function was not stabilizing. Each set of labs told the same story slightly more clearly than the one before it. What had been described as worth monitoring was now being described in terms that required closer attention. Nashville had offered distance. It had not offered answers.

He left.

He went back to Wisconsin.

Not because it had been kind to him, but because it was known. His siblings were there. People who had his history, who did not require explanation, who could simply be present without needing him to construct context. At this stage, known was its own kind of medicine.

"The South reminded me too much of Mississippi," he said. "The racism down there wasn't subtle. It was right in your face. I had already seen that kind of thing. I didn't need to keep putting myself in it."

When he settled back in, the medical monitoring intensified. What had been periodic appointments became regular ones, and what had been cautious language from his doctors became more direct. Something in his kidney function warranted a different kind of conversation. His doctors began asking different questions.

Specific questions.

"He asked me about my military history," Lewis told me. "Where I had been stationed."

Standard intake at a certain point, particularly at the VA. Lewis had answered it before. He answered it the way he answered everything—directly, completely, in the order it happened.

Training. Vietnam. Morocco. Wiesbaden. And then, Camp Lejeune.

“He stopped when I said that,” Lewis told me. “He looked at me different.”

The doctor’s pen came down from the pad. His posture changed. There is a particular quality to the moment when a medical professional suddenly understands the patient in front of them in a way they did not moments before. It is not always visible. With this doctor, it was.

“He said, ‘Camp Lejeune?’” Lewis recalled. “And I said, ‘Yes.’”

The doctor began to explain.

What he said to Lewis that afternoon Lewis had not known. Not the specifics. Not the timeline. Not the chemicals or the dates or the scope of it.

The water at Camp Lejeune had been contaminated with industrial solvents since at least 1953. Trichloroethylene. Perchloroethylene. Benzene. Vinyl chloride. They had entered the groundwater from dry-cleaning operations, degreasing facilities, fuel storage, the infrastructure of a functioning military base, leaking quietly into the earth, and from the earth into the pipes, and from the pipes into the taps. Into the canteens. Into the food. Into the bodies of every person living and working and training and being born on that installation.

Government testing in 1982 had confirmed contamination at concentrations hundreds of times above safe limits. In some wells, thousands of times. The disclosure did not come until 1987. The water ran for five years after they knew.

Lewis's wife had been pregnant on that base. Lewis had filled his canteen from those pipes every morning. He had brushed his teeth with that water. He had showered in it, cooked with it, consumed it as the most basic assumption of daily life—that the water your own government gives you is safe to drink.

The kidney disease that had refused to respond to treatment regardless of geography. The blood pressure that had resisted correction across three states. The fatigue that no amount of discipline or medication could fully address. The declining function that labs had been recording and doctors had been monitoring without being able to explain what was driving it.

It had a source.

It had always had a source.

Lewis sat with the doctor's explanation moving through him the way cold water moves—not all at once, but finding every available space. He thought about Wisconsin. About Savannah. About the miles he had put between himself and the place he believed was making him sick. The fact that the illness had traveled with him faithfully through all of it because it was not in Wisconsin, was not in the air or the tension or the accumulated pressure of decades of injustice.

It was in him.

It had been in him since 1967.

"All that time," he said to me, very quietly, "I thought I was running from what I had been through."

He let a beat of silence land between us.

"But it was already in me."

He looked at me then—not with anguish, and not with the performed composure of a man who will not allow himself to feel what he is saying. He looked at me the way someone looks when the last piece of a long equation has finally resolved and the answer, though devastating, is at least legible.

"I was poisoned twice," he said.

He did not raise his voice. He did not need to.

"Once in Vietnam," he continued. "And once at Camp Lejeune."

The first time, a war had done it, Agent Orange drifting down through the jungle canopy while he moved underneath it, absorbed into skin and lungs during months of exposure that no one told him to be concerned about because no one in authority was prepared to say out loud what the planes were doing to the men on the ground. The second time, his own government had done it in peacetime, on American soil, in a base where he had reported as ordered, and they had watched the results of their own water testing for five years and said nothing.

He was not a man who had neglected his health. He was not a man who had ignored his body or treated it carelessly or failed to do what was asked of him when doctors gave instructions. He had done everything right. He had been disciplined about his body since the day the Corps taught him how to inhabit it, and he had not stopped.

The discipline had not been the problem.

The water had been the problem.

And the water had been inside him long before he understood what was wrong.

Chapter 10: They Are Waiting for Us to Die

"I didn't need an enemy to shoot at me. The enemy was already in the water we drank every day."—Lewis Cain

---

He sat with it for a moment after he said that. The words hung between us in the kitchen like smoke from something that had been burning for a very long time, and I did not rush to fill the space. I had learned, by then, that Lewis's silences were not empty. They were full. They held things that language was still catching up to.

We had been sitting at that table for hours. The recorder was still running. My coffee had gone cold, and Lewis's had gone colder, and neither of us had moved to warm either cup, because to move felt like it might break something fragile that was finally coming through.

"Poisoned twice," he said again. Not louder. Not with anger. Just the way a man says a thing when he needs to make sure it was heard.

"Once in Vietnam. Agent Orange dropping out of the sky while I was trying to stay alive in a jungle they sent me to. And once at Camp Lejeune. By the water they handed me in a cup and told me to trust."

I looked at him. At the way his hands rested on the table, steady and deliberate. At the way his jaw set when the truth of it pressed against him from the inside. I thought about everything he had told me in the hours before that sentence arrived. Mississippi. The dogs. The hospital bed. Vietnam. Reed falling in the rain with his hands at his throat. Morocco. Wiesbaden. Lakeside County and the decades of paperwork that followed him like a shadow he could not outrun.

And now this. Camp Lejeune. The water.

I asked him to take me back there. Not to the courtroom. Not yet. Back to the beginning of it, the moment the base became part of his body without his knowing.

He exhaled slowly through his nose.

“All right,” he said. “Let me take you there.”

The faucet at Camp Lejeune sputtered before the stream steadied.

That was the first thing he told me. Not the sounds of drills or the weight of his pack or the heat that pressed down on North Carolina like a hand on the back of your neck. The faucet. The way it hesitated before it found its flow, the way the pipes groaned like old men clearing their throats in the morning. The way it felt routine.

“It was just water,” he said. “You didn't think about it. You didn't question it. You filled your canteen, you drank it, you moved on. That was it.”

He was twenty-two years old. He had already survived Vietnam. He had already lost Henry to the river and Reed to the rain, and he had watched things happen in that jungle that he was still carrying fifty years later. He had survived Morocco and the bottle and Wiesbaden and the long slow process of being moved from place to place by a military that was trying to figure out what to do with a Marine who had broken under the heaviness of a war that should never have been asked of him. He had survived all of that and arrived at Camp Lejeune believing the worst was behind him.

The base hummed with the rhythm of preparation. Boots struck pavement in cadence. Drill instructors' voices cut through the thick Carolina air. Young men filled canteens and aluminum pitchers and tin cups from the same water, morning after morning, laughing through the fatigue, brushing their teeth in sinks that ran off the same pipes, showering under the same pressure, cooking in kitchens fed by the same underground veins.

Nobody told them.

Not in 1967, when Lewis arrived. Not in the years that followed. The water ran clear. It had no smell. It offered no warning. It was simply there, the way everything provided by the military was simply there, because you trusted the system that gave it to you, the same way you trusted the orders that sent you somewhere, the same way you trusted that the uniform on your back meant something and that the country inside it would honor what it had asked you to give.

“We didn't question it,” Lewis said again, and I could hear the weightiness that had settled onto those words in the decades since. “You don't question the tap your own government gives you. That's not naïve. That's just how it was.”

His son was born at the base hospital. December of 1967, at the Naval Hospital on the same base, fed by the same water. His wife had carried that baby on a base they had been ordered to live on, drinking what they were given, trusting what they were told. When the baby came, he drew his first breath inside the same walls, the same system, the same contamination that was already moving through the ground beneath all of them.

Nobody told them then either.

They would not be told for another two decades.

The contamination had been there since at least 1953. Government documents, eventually unearthed after years of pressure and litigation, would confirm it. The wells at Tarawa Terrace and Hadnot Point, two of the main water systems feeding the base, had been saturated with industrial chemicals for decades. Trichloroethylene, used to degrease military equipment, had seeped from maintenance operations into the groundwater. Perchloroethylene from dry cleaning facilities bled into the same underground system. Benzene from fuel storage tanks leaked slowly downward. Vinyl chloride formed as those compounds broke down in the earth and found their way into the pipes.

The levels were not trace amounts. They were not borderline. Internal testing conducted in 1982 showed contamination at concentrations hundreds of times higher than what would be considered safe today, some readings reaching thousands of times the acceptable limit. Those results were documented. They were understood. They sat inside the system that was responsible for protecting the people on that base.

And for five years, the water continued to run.

For five years after the government's own testing confirmed what was in the wells, Marines filled their canteens, families bathed their children, wives washed bottles for infants in kitchen sinks, and nobody said a word. The disclosure did not come until 1987. By then, the damage had already been done. By then, the chemicals had already traveled through hundreds of thousands of bodies and begun their slow, invisible work.

When Lewis described learning this for the first time, sitting at his kitchen table with documents spread out in front of him, his voice carried something I had not heard before in all our hours together. It was not anger, though anger would have made sense. It was something quieter and harder. The sound of a man who has done his accounting and found the numbers do not lie.

"Nineteen fifty-three to nineteen eighty-seven," he said. "That's thirty-four years. And my son was born there in sixty-seven. My wife was there. I was there. And they knew. They had it on paper. And they didn't say anything."

He pressed his palm flat on the table the way a man anchors himself when the ground beneath him has shifted, and he needs to know something solid is still underneath him.

"You want to know what they were waiting for?" he asked.

He answered before I could respond.

"For us to die. That's what I think. If enough of us die before we figure it out, you don't have to explain yourself. You just wait."

He had appeared on the local news before the court date. A television reporter in Lakeside had reached out after hearing about his case, and Lewis had sat under studio lights with another veteran at his side and said exactly what he thought in exactly the plain language he always used.

"When did you learn what was in the water?" the reporter had asked.

"Decades later," Lewis told her. "Back then it looked clean. Tasted like water. We trusted it because it came from the same people who trained us."

She had asked him what he wanted viewers to understand.

"We served this country," he said. "We trusted this country. We're not asking for gifts. We're asking for honesty."

And when she asked what he felt about the pace of the legal process, he looked straight into the camera and said the thing he had been thinking for a long time.

“I feel like they are waiting for us to die. If we die first, they don't have to pay us. Or admit what they did.”

No anger in his voice. No performance. Just a man who had been fighting long enough to have arrived at a truth he no longer felt the need to soften.

I asked him, sitting at his kitchen table, if he believed that.

“I know it,” he said. “I can count. I can read a calendar. These men are in their seventies and eighties. Every year that goes by, more of them are gone. And the cases don't move any faster when they go. You just have fewer claimants to deal with. That's not a coincidence.”

The federal courtroom in North Carolina was not what you see in movies. There were no dramatic entrances, no gasps from the gallery, no music swelling underneath the significance of the moment. What there was, was quiet. The kind of quiet that happens when a large room is full of people who are already exhausted and have been exhausted for a very long time.

Lewis described it to me in the way he described everything, with precision, with detail, with the careful attention of a man who has learned that what you notice matters.

They came from across the country. Men in their seventies and eighties, some older. They moved the way men move when their bodies have been carrying more than they were designed to carry, slowly, deliberately, aware of every joint and every breath. Some leaned on canes that tapped softly against the tile floor in a rhythm that sounded almost like a metronome. Others sat in wheelchairs pushed by sons or daughters who had driven through the night to be there. Several were connected to oxygen machines that hummed at low volume and reminded everyone in the room with every breath that time was the one thing they did not have in abundance.

"It looked like a formation," Lewis told me. "That was the first thing I thought when I walked in. These men know how to sit in rows. They know how to be still when it matters. It looked like a second roll call."

He took his seat with the binder beside him, its vinyl spine worn from years of handling, its pages organized into tabbed sections with the same discipline he had brought to everything since Vietnam. Medical records in one section, legal correspondence in another. Reports with dates underlined and passages circled in ink that had not faded because Lewis Cain did not use cheap pens when the documentation mattered. His mother's Bible was in the inside pocket of his coat, as it had been in every courtroom, every hearing, every waiting room he had sat in for thirty-five years.

He pressed his hand against the binder and felt the power of it.

“All those years,” he told me. “All that paper. And it came down to sitting in a room with men I hadn't seen since we were boys and making sure somebody with authority had to listen.”

A veteran named Miller was called to testify before Lewis. His hands trembled when he adjusted the microphone, but his voice did not tremble at all.

“I was stationed at Hadnot Point,” Miller said. “Nineteen sixty-six to sixty-eight. We drank that water. Showered in it. Brushed our teeth with it. Nobody warned us.”

The government attorney did not look up immediately. He finished a note on his legal pad, adjusted his glasses, and then posed his question in the measured tone of a man who had been taught that precision is a weapon.

“Mr. Miller, can you say with certainty that your current condition was caused directly by water consumed at this base nearly sixty years ago?”

The gallery shifted. Lewis felt it, the collective tightening of men who recognized the maneuver. It was the same maneuver that had been used against all of them in different rooms for different cases over different decades. The careful clinical phrasing designed to place the burden on a man's memory and biology rather than on the documented record.

Miller leaned into the microphone.

"I can say with certainty I was twenty years old when I got there," he said. "Healthy as a horse. I can say with certainty I buried three men from my unit who died of the same thing. And I can say with certainty your own reports show benzene and TCE in those wells at levels so high you would shut down a factory today."

The attorney pressed. "But you cannot definitively connect your diagnosis to that specific water source. Are there not other risk factors that—"

Miller's hand came down on the table.

The crack of it went through the room like a rifle shot, and then settled into a silence that was absolute.

"I did not smoke," Miller said, eyes locked on the attorney. "I did not drink. I served. When I came home, the only thing in my blood besides grit and God was what you put in that water."

The judge tapped her gavel once, a reminder, not a reprimand.

The attorney sat back. But Miller's words did not sit back with him. They stayed in the air.

Lewis sat in the gallery and felt the words settle into his chest the same way cold water settles when you swallow it quickly. He looked at the men around him. Faces he did not know by name, but recognized by damage. The dialysis pallor. The bone-tired eyes. The way certain men held themselves upright with a discipline that had no reasonable explanation except that Marines do not know how to stop standing at attention even when their bodies are begging them to stop.

More veterans rose to speak.

A man named Magee, thin as paper and steady as stone, leaned into the microphone and said, simply, "People were dying every day. So, it became commonplace." He did not raise his voice; he did not need to. The plainness of his truth carried its own power.

Another Marine rose with a cane that clattered against the table as he braced himself upright. "We thought it was just us," he said, scanning the gallery. "We all thought it was just us. But it was all of us. All along."

His voice cracked, but he did not stop.

Lewis told me that at some point during those testimonies the room stopped feeling like a courtroom and started feeling like something else entirely.

"Like a church," he said. "Like testimony. Like every one of those men was bearing witness to something that the rest of the world had been trying not to see."

When Lewis was called, he rose slowly. He adjusted the binder under his arm. He moved to the witness stand with the deliberate economy of a man who has learned that stillness communicates more than motion. He sat, placed the binder on the ledge in front of him, looked at the veterans' attorney, and waited.

"Mr. Cain," the attorney began, "can you tell this court where your service brought you?"

"Camp Lejeune, North Carolina," Lewis answered. "Nineteen sixty-seven."

"Describe that time for us."

Lewis let the silence breathe for just a moment before he answered. Not hesitation. Preparation. The way a man gathers himself before he says something that deserves to be said correctly.

"Early mornings," he said. "Boots hitting pavement together. Drill instructors shouting. Sweat, gun oil, salt air off the coast. We ran until our lungs burned. We learned to crawl low and move fast. That base was where boys got broken down and built back as Marines."

"And the water?"

His eyes moved to a point just above the attorney's head for a moment. A gesture I had seen before at his kitchen table when the memory was coming from a distance.

"The faucet sputtered," he said. "Then it flowed. Looked clear. Had no smell. We drank it straight from the tap. We filled canteens with it. Showered in it and cooked with it. My wife drank it. My son was born at the hospital on that base. Nobody told us there was anything wrong with it."

"And did you have any reason to believe it was unsafe?"

"No," Lewis said. "We trusted it because we trusted the people who gave it to us. That's not something I'm ashamed of. That's what trust is."

The attorney moved closer.

"Years later, what did you learn about that water?"

Lewis leaned forward slightly. The binder was in front of him. He did not open it. He did not need to. He had read it enough times that the documents had become part of his own memory.

"I learned it was full of chemicals. Trichloroethylene. Perchloroethylene. Benzene. Vinyl chloride. Stuff from dry cleaners, fuel tanks, and degreasing operations. Your own tests in nineteen eighty-two showed levels two hundred, three hundred, some readings three thousand times higher than what is considered safe now. We didn't sip it once. We drank it every day."

A murmur moved through the gallery. Lewis did not look back at it. He kept his eyes forward.

"And how has that touched your life?" the attorney asked.

Lewis's jaw worked for a moment.

"Kidney disease," he said. "Pain that does not leave. My ex-wife has the same kidney disease. My son, born at the base hospital, has fought a skin condition his whole life that no doctor has been able to cure."

He paused.

"Some things," he said, "are harder to say out loud than others."

The room understood that.

The attorney let the silence do its work, then asked his final question.

"Why are you here today, Mr. Cain?"

Lewis sat straighter. His eyes were clear and steady and burning with something that had not dimmed in eighty years.

"Because this is not just my story," he said. "It belongs to the men in this room. It belongs to men already buried. They trained us to dodge bullets. They never told us we'd need to dodge the water. Now they want to wait us out with silence and delay until we're gone. I'm still here. And I kept the record."

He tapped the binder once. The sound was small.

In the quiet of that courtroom, it landed like a hammer.

The government attorney stood for cross-examination with the composure of a man who had been trained never to show when a room has moved against him.

"Mr. Cain," he began, "you testified that you drank Camp Lejeune water in nineteen sixty-seven and sixty-eight. At that time, you had no medical diagnosis related to that water. Correct?"

"Correct," Lewis said. "Poison does not always show its hand the next day. It can take years. Sometimes decades. Like rust in a pipe you don't see until it bursts."

"But you cannot prove that your kidney disease was caused specifically by that water."

Lewis leaned a fraction forward.

"I can prove I was there," he said. "I can prove your own tests in nineteen eighty-two showed chemicals in those wells up to three thousand four hundred times higher than what is safe, and nobody told us until nineteen eighty-seven. I can prove thousands of Marines and family members who lived on those water lines developed the same kinds of illnesses. You can give it any name you want. I call it cause and effect."

A current moved through the veterans' gallery. Cane handles tightened. Throats swallowed. Nobody made a sound.

The attorney tried again.

"Mr. Cain, you are eighty years old. You have other health conditions that can simply come with age. Is that fair to say?"

"Yes," Lewis said. "I earned every one of these years. My ex-wife has the same kidney disease. My son has lived his entire life with a skin condition he was born into while I was stationed at that base. You want to tell me age explains that too?"

The room stilled. The air moved. No one spoke.

The attorney shuffled his papers. Tried once more.

"You have spent much of your adult life filing complaints and appeals," he said. "Some might characterize that as seeing wrongdoing where none exists."

Lewis looked at him without blinking.

"You can call it whatever you like," he said. "I call it paper. Signed reports. Missing files. Contradictions in black and white. I have kept every single one of them. If speaking the truth year after year makes me persistent, I'll wear that the same way I wore my medals. They came with a cost. So does this."

The attorney's jaw flexed. He tried one last question.

"Do you honestly believe this court owes you anything more than sympathy?"

Lewis did not blink. His voice dropped lower and steadier, the way a man's voice drops when he has finally arrived at the center of the thing he has been circling for a lifetime.

"I did not come here looking for sympathy," he said. "I came here for accountability. You let us drink poisoned water, then hid the truth and waited while men died. Sympathy does not fix that. Justice might."

The silence that followed was not empty.

It was full of thirty-four years.

The judge sat still longer than the rules required after Lewis returned to his seat.

She was a woman of careful eyes and the practiced stillness of someone who had heard a great many things in courtrooms and had learned to let the importance of what she'd heard settle before she moved. Her pen hovered over her pad without touching it. The government attorney shuffled papers at his table. Nobody spoke.

Lewis sat down among his brothers.

And then it started.

One cane. Tapping the floor.

Then, another.

Then, another after that.

Not applause. Not cheering. Something older than both. A rhythm that started at the back of the gallery and moved forward through the rows of men who had once stood in formation on a base in North Carolina before the world did to them what it did. Canes lifted and came down in a cadence that was not quite military and not quite music, but was something between both, the sound of men who still knew how to move together even when their bodies had been broken by the same betrayal.

The judge did not stop it.

She tapped her gavel once, gently, the way you tap something when you want it recorded in the room rather than ended.

Lewis looked at the binder on the ledge beside him. He looked at the men tapping their canes. He thought about Reed. He thought about Henry. He thought about his mother praying in the early morning dark of a Mississippi kitchen, talking to God the way you talk to someone sitting right across the table from you.

"She used to say," Lewis told me, back at his kitchen table, voice just above a whisper, "'The Lord didn't bring you here to leave you here, Lewis.'"

He looked at me.

"I'm still here," he said. "After everything. I'm still here."

I could not answer him right away.

So, I sat with it, the same way the courtroom had sat with the canes, the same way he had been sitting with the truth of his own life for eighty years, letting it be present without trying to resolve it into something smaller than it was.

Chapter 11: I Will Not Be Erased

"I don't need money in heaven. I don't think they've got banks up there. But people down here are going to need it. So, I want the truth on the record before I go."— Lewis Cain

The afternoon had fully arrived by the time Lewis said what came next.

The kitchen light had changed. The recorder between us on the table had been running for so long that I had stopped being aware of it, which is the way it should go in an interview like this one, when the person across from you has finally stopped speaking for the machine and started speaking for themselves. The coffee was long gone. The questions I had prepared were somewhere at the bottom of my notebook, buried under hours of things I had not known to ask because I had not known enough to ask them.

Lewis had been quiet for a moment. Not the silence of a man who has run out of things to say, but the silence of a man deciding where the ending of a thing should be. And when he spoke again, he spoke the way he had spoken about his mother, about Vietnam, about Reed, about Henry, about the water and the binder and the thirty-five years. Simply. Without decoration. As if the truth he was carrying had been waiting all this time for a room quiet enough to speak in.

"I don't have much time left," he said.

He said it the way a man says a weather report. Factually. Without theatrics. Without asking anyone to feel sorry about it.

"I'm not complaining about that. God gave me eighty years, and I used every one of them. I used them fighting, which is maybe not how most people would choose to spend a life, but that's how mine went. And I made my peace with it a long time ago because the fighting wasn't really about me. It was about what was right. It was about making sure that when I was gone, the record said the truth."

I looked at him across the table. At the man he had become through everything he had told me. At the boy who couldn't walk in a Mississippi hospital bed holding a red fire truck. At the young recruit running through sand in San Diego until he outran the whole platoon. At the Marine pressing his hands against Reed's throat in the dark jungle rain, saying, please, please stay with me, and at the man with the binder and the Bible, walking into courtrooms for thirty-five years.

"What do you want people to know?" I asked him. "When someone reads this book. When someone watches the movie, what do you want them to know?"

He did not answer immediately. He looked out toward the window where the afternoon light was softening toward evening.

"I want them to know that I was here," he said. "That's the first thing. I was here, and I did not go quietly. I want them to know that when people tell you what you can't do, that's not information about you. That's information about them. That doctor in Mississippi told my mother I'd probably never hold a job. I worked for Lakeside County for thirteen years. I served my country in a war that most people have already stopped thinking about."

He turned back to me.

"I want them to know that the system is built by people, and what people build, people can change. But only if you are willing to do what I did. Keep the paper. Keep the record. Don't let them tell you that what happened didn't happen. You were there. You know what happened. And if you have the documentation, they cannot take the truth away from you, no matter how hard they try."

He reached into the inside pocket of his jacket and placed his mother's Bible on the table between us.

It was small. Worn at the corners, the leather cover softened by decades of hands. The spine was cracked the way a book gets cracked when it has been opened at the same passages so many times that the binding has learned where to land on its own. There were slips of paper tucked between certain pages, and on the front inside cover, written in a hand that was not Lewis's, a name I could not quite read from where I was sitting.

“She gave this to me before I left for Vietnam,” he said. “I carried it in my pack through the whole tour. I had it in Morocco. I had it at Wiesbaden. It's been in every courtroom I've ever walked into.”

He looked at it on the table.

“She used to say, 'Lewis, God did not bring you here to leave you here.' Every time things got bad. Every time the ground felt like it was going out from under me. She'd say it like she was reminding God of something He'd promised her personally.”

He smiled at that. A quiet, private smile, the kind that belongs to a memory too specific to explain, but too good not to carry.

“I think she was right,” he said. “I think about everything that tried to kill me and didn't. The rheumatic fever. The dogs in Mississippi. Vietnam, three tours in a jungle where men who were stronger and faster and luckier than me died all around me. The bees that stopped my heart. Morocco. Agent Orange. Camp Lejeune and that water. Thirty-five years of a system doing everything it could to make me disappear. And I'm still here. At eighty years old, I'm still here, still fighting, still talking.”

“That's not luck,” he said.

“No,” I agreed.

“That's purpose,” he said.

He told me about his morning.

Every morning, before the day began, before the pain made itself known fully, before the phone calls or the correspondence or the legal files, he made his coffee. He sat with it. And he talked.

“I talk to God the way my mama did,” he said. “Like He's right there. Like we're just two people at a table, and I'm telling Him what I need and what I'm grateful for and where I'm at. I don't make it complicated. I don't use a lot of church words. I just tell him the truth.”

“What do you tell Him?”

“I tell Him thank you,” Lewis said. “Every morning, first thing. Because I'm still breathing. Because the kidneys are still working, even if they're not working as they should. Because I have the binder. Because I have all this documentation that they tried so hard to bury and it's still here, same as I am.”

He picked up the Bible from the table and held it for a moment.

“I tell Him I want the case resolved before I go. Not for me. I want it for the people who come after me. I want it for the veteran who's sitting in some waiting room right now getting told that his disability rating doesn't match what the doctors say. I want it for the Marine who's going to find out twenty years from now that the water at the base where she trained was contaminated and nobody told her. I want it so that what happened to me is on the record, not hidden in a filing cabinet somewhere. On the record. Where it can't be undone."

"And if the case isn't resolved?"

He did not hesitate.

"Then, the book is the record," he said. "That's why I contacted you. That's why God brought you to me. Because the court system has been trying to wait me out for thirty-five years and I needed a different kind of record. A permanent one. One that doesn't get dismissed and doesn't get rescheduled and doesn't forget what it knows."

He looked at me with the steady, clear eyes of a man who has lived long enough to understand exactly what he is saying and exactly why it matters.

"That's this book," he said. "You're the record."

There is a photograph on the wall of Lewis's living room. I noticed it when I first walked in, before we sat down, before the recorder was running. A young man in a Marine Corps uniform, standing straight with the particular posture that the Corps builds into you and never entirely takes back. Eyes forward. Shoulders squared. The expression on his face was a mix of pride and readiness, that of a man who has been told he is part of something larger than himself and believes it.

I stood in front of it for a moment without saying anything.

Lewis came to stand beside me.

"That's me," he said. "Nineteen sixty-five."

The year he shipped to Vietnam. The year the first ground troops landed on a beach in a country most Americans could not find on a map. The year a boy from Jackson, Mississippi who had survived rheumatic fever and Mississippi dogs and threats and poverty and the long slow war of growing up Black in a state that considered his existence a problem to be managed, put on a uniform and went to fight for a country that had not finished deciding if it would fight for him.

He had gone anyway.

That is the part that stays with me. Not the legal cases, though they matter. Not the contamination, though it is unconscionable. Not even the forged documents, the administrative betrayals, or the decades of delay, though all of it is real and all of it demands to be on the record.

What stays with me is simpler than all of that.

He went anyway.

He served anyway.

He trusted anyway.

And when they broke that trust, he did not go quietly. He opened the binder. He wrote the letters. He filed the complaints. He showed up to the hearings alone when no attorney would stand beside him. He sat before judges, administrative law judges, reporters, and eventually a writer at a kitchen table. He said, plainly and without flinching, "This is what happened, and here is the documentation, and I will not be erased."

The last thing Lewis said to me that day was not about the case. It was not about the federal court, the motion for summary judgment, or any of the procedural machinery that was still turning.

It was about money.

He said it the way his mother had taught him. Without attachment to it. With a clarity about what it was for and what it was not for.

"I don't need money in heaven," he said. "I don't think they've got banks up there. But people down here are going to need it. The veterans. The young Black men who don't know their rights. The families that are going to find out twenty years from now that the base where their father trained was poisoned and nobody told them. I want whatever comes out of this to go to them."

He looked at the Bible on the table.

"The money was never the point," he said. "The point is justice. And justice doesn't just mean you get paid. Justice means the record says the truth. Justice means the next person who sits in a waiting room getting told their case is closed has somewhere to point and say, this happened before, here's how it went, here's what you do, here's proof that the fight is worth fighting."

He looked at me.

"That's the book," he said. "That's what this is."

I told him I understood.

He nodded once. The way a Marine nods when something has been confirmed; and the matter is settled; and the mission continues.

"Good," he said. "Then, let's make sure we get it right."

I drove home as the evening came down over the highway and I thought about what Lewis Cain had survived.

I thought about his mother's hands on his legs, day after day, willing them back to life. I thought about him running through the San Diego sand until he outran everyone. I thought about the jungle and the river and Reed and Henry and the bees and Morocco and the long plane ride to Wiesbaden with a man on each side who wouldn't speak to him. I thought about Lakeside County, the binder, and the decades of paperwork. I thought about Camp Lejeune and the water that ran clear and the thirty-four years of silence and the kidneys that had been failing slowly for decades without anyone knowing what had put them on that path.

And I thought about the photograph on his wall. The young Marine in that photograph. The squared shoulders. The eyes that had not yet seen what they were about to see.

He did not know that day in nineteen sixty-five what was coming. He could not have known. He was a boy from Mississippi who had survived enough already to believe that survival was its own form of purpose.

He was right about that.

He has been right about most things.

And he is still here.

At eighty years old, kidneys failing, eyes clouded, body marked by every war that was fought in it and every system that tried to grind it down, he is still here. Still at the kitchen table every morning with his coffee. Still talking to God the way his mother taught him to. Still keeping the binder and still pressing for the case that has been pending for longer than some lawyers have been alive.

Still in formation.

Whatever the courts ultimately decide, whatever the system ultimately does or refuses to do, one thing is already settled. The record exists now.

He was here.

He fought.

And he would not be erased.

Chapter 12: What Witnessing Costs

"You see why God picked you for this position?"— Lewis Cain, to Dr. Kara Lock-Harris

---

The last interview ended the way the first one began. Quietly. With coffee going cold on the table between us and something sacred in the air that neither of us tried to name.

I drove home that evening with the recorder still warm in my bag. Lewis' voice still moving through me, not fading the way voices usually do when you leave a room, but settling, the way something settles when it has finally found the place it was always meant to rest.

I had crossed the threshold of his home not knowing what I was walking into. That part belongs to Chapter 1. What belongs here, at the end, is what I carried out.

Writing is craft. You can learn it, practice it, improve it. You can prepare for it the way I prepared for our first meeting, notebook open, questions typed in careful order, a plan. But witnessing is something else. Witnessing is what happens when the plan dissolves, and the heaviness of another person's truth becomes something you can feel in your own body; witnessing costs you something. It rearranges you. And you do not always know what has shifted until you are driving home in the dark, replaying a sentence Lewis said three hours ago that you only half understood when he said it, and suddenly it opens.

I did not come into this knowing how to carry that.

Lewis taught me.

Over the hours we spent together at his kitchen table, I learned what that cost looked like in a life. I saw it in the binder worn soft at the spine from years of handling. I saw it in the way he described Reed dying in the rain without his voice breaking; not because the grief was gone, but because he had learned to carry it while still breathing. I saw it in the exercise bike positioned in the center of his living room, like an act of defiance against a body that had been given every reason to stop. I saw it in the photograph of a young Marine standing straight in a uniform, eyes forward, not yet knowing that the fight of his life had not started yet.

Witnessing means staying in your seat even when the burden of what you are hearing becomes something you can feel in your own chest. It means understanding that what is being given to you is not entertainment and not an argument, but evidence. Evidence that a man lived. Evidence that he fought. Evidence that the system tried to erase him, and he would not be erased.

I had launched my business two years before his message arrived. I had been building it with the quiet persistence of someone who believes in what they are doing but cannot yet see the shape of where it is going.

There were days when I understood exactly why I had started, and days when the understanding lived somewhere just out of reach, present but not visible, the way faith sometimes operates when you are in the middle of a stretch that has not yet shown you its purpose. I had prayed not for success in the way success is usually defined, but to be pointed toward the work that was meant for me. I had felt, more than once, that I was supposed to wait. That the right moment would come with a clarity that would make the waiting make sense.

Then Lewis Cain sent a Facebook message on a Sunday morning, and I almost did not open it.

That is the part I return to most often when I think about how this book came to be. The almost. The nearness of missing it. The way God sometimes places the most important thing directly in your path and still requires you to choose to pick it up. I chose. And in choosing, I found not just a story to tell but a mirror to look into. Because sitting across from Lewis, hearing what he had survived, watching how he carried it, I saw something I needed to see about my own capacity for endurance. About what it looks like to fight for thirty-five years without surrendering the thing inside you that knows why you started. About what faith looks like when it is not a feeling but a practice—the same practice every morning, coffee and conversation and gratitude before the day begins, no matter what the day contains.

I needed the example of him as much as he needed the witness of me.

That is what God knew when He brought us to the same table.

The legal cases are ongoing as this book goes to press.

The Camp Lejeune claims remain in process, along with the claims of hundreds of thousands of other veterans and family members who drank from the same contaminated water. The Lakeside County workers' compensation case, built on decades of documentation and the discovery of settlement documents that Lewis contends were never agreed to, continues before the courts. Lewis represents himself. He continues to press for a ruling on the question he has been asking for 35 years: whether a dismissal obtained through documents he never signed is valid.

He believes the answer is no.

The names of all government officials, attorneys, and county personnel referenced in this book have been changed or withheld to protect the integrity of ongoing legal proceedings. All events, timelines, medical records, and legal disputes described are drawn from Lewis's personal archives and from the hours of interviews we conducted together across multiple sessions. Nothing in these pages has been fabricated. Everything Lewis told me, he told me with his documentation open on the table between us.

He kept the record. This book is now part of that record.

The morning after our last interview, I sat at my kitchen table with my Bible and coffee. The steam curled upward the way it always does. The house was quiet in the particular way it is quiet before the world arrives. And I prayed for Lewis.

I prayed for his kidneys, which carry what the water gave them and keep going anyway. I prayed for his son, who carries a skin condition he was born into on a contaminated base. I prayed for the thousands of veterans who are still waiting, whose bodies have been keeping count of what was done to them since before they understood what was being done. I prayed for the ones who did not live to see this moment, whose stories went to the ground with them before anyone thought to write them down.

And I prayed for his mother. Mary Lindsay. The woman who built a house with her own hands, raised thirteen children in it, sent her boys north to keep them alive, put her hands on a sick child's legs day after day, and refused to accept the doctor's prediction, prayed in her recliner every morning before the sun came up, talking to God like He was sitting right there at the table with her.

I think He was.

I think He is still with Lewis.

I think that is why Lewis Cain is still here.

"You see why God picked you for this position?"

That is what Lewis said to me early in our relationship, before the interviews fully began, before the chapters took shape, before any of this became what it has become. He said it with the confidence of a man who does not say things he does not believe.

I did not know how to answer him then.

I think I know now.

God chose me for this position because Lewis's story is not about one man. It is a story about what happens when the systems that are supposed to protect people decide to protect themselves instead. It is a story that belongs to every person who has ever been told their word is not enough, their pain is not provable, their documentation is insufficient, and their persistence is excessive.

It is a story for everyone who has been standing in a room where everyone can see the fire, but nobody wants to say it is burning.

You are not wrong about what you see.

Open the binder.

Keep the record.

Stay in formation.

Lewis Cain did. And after eighty years, after everything this country asked of him and everything it took from him, he is still here. Still fighting. Still pressing for the truth to be written where even the system cannot erase it.

This book is that place.

And now, so are you.

Acknowledgments

This book exists because Lewis Cain trusted me with the most consequential story of his life. He handed me his binder, opened his records, gave me hours of recorded conversation, and then did the harder thing—he waited while I found the words to hold it. Mr. Cain, this book is yours. I was only given the privilege of writing it down.

To Dr. Vanessa R. Liederbach, whose editorial eye is as precise as it is generous. Thank you for reading every chapter with the kind of care that made this manuscript better than I could have made it alone. Your feedback was not just technical. It was faithful to the story.

To my son, James, who wears the uniform today, I wrote this book thinking of you and the sacrifice you make. Every page carries a prayer for your protection, your purpose, and your safety. May the men who come after Lewis Cain find a system worthy of the sacrifice they are making. That is why this story had to be told.

To every veteran who has ever sat in a waiting room, filled out a form, been told no, and filled it out again, this book was written with you in mind. Your persistence is not stubbornness. It is the same discipline that carried you through the hardest days of your service. Keep the record.

And to my Lord and Savior Jesus Christ, who placed this assignment in my path before I knew to look for it, thank You for the cup of coffee, the open Bible, and the Sunday morning message that found me at exactly the right moment. I am still in awe of Your precision.

Author's Note

Battle Hymn of the Disregarded is a memoir grounded in truth and shaped by the tools of narrative craft.

The events, timeline, and circumstances described in this book are drawn from multiple recorded interview sessions conducted with Lewis Cain over the course of more than a year. Lewis's personal documentation—decades of medical records, legal correspondence, workers' compensation filings, pension notices, hearing transcripts, and the contents of the binder that appears throughout these pages—served as the primary evidentiary foundation for the narrative.

Dialogue throughout the book reflects Lewis's direct accounts as shared in recorded conversation. Where scenes and exchanges are reconstructed, they are built from his testimony and presented in the spirit of what he described, not invented beyond the bounds of what he told me.

To protect the integrity of ongoing legal proceedings and the privacy of individuals involved, certain names of government officials, attorneys, law enforcement personnel, and county employees have been changed or withheld. Some locations and identifying details have been adjusted for the same purpose. These changes do not alter the substance of the events described.

The contamination of the water supply at Camp Lejeune, the chemicals involved, the dates of exposure, and the government's own internal testing and disclosure timeline are matters of public record, documented in federal litigation, congressional testimony, and the Camp Lejeune Justice Act of 2022. All factual claims related to the contamination are drawn from these public sources.

Lewis Cain's legal cases are ongoing as of the date of this publication. Nothing in this book should be construed as legal advice or as a complete account of any pending matter.

I wrote this book as a witness. The record belongs to Lewis.

Resources

If you or someone you love was stationed at Camp Lejeune between 1953 and 1987, or if you are a veteran navigating a disability claim, benefits appeal, or service-connected health condition, the following resources may be of assistance.

Camp Lejeune Justice Act (CLJA)

Signed into law in August 2022, the Camp Lejeune Justice Act allows veterans, family members, and civilian workers who were exposed to the contaminated water at Camp Lejeune to file claims for harm. Information is available through the U.S. Department of Justice and the Department of Veterans Affairs.

www.va.gov/camplejeune

U.S. Department of Veterans Affairs (VA)

The VA provides healthcare, disability compensation, pension, and benefits assistance to eligible veterans and their families. Veterans can file claims online, by phone, or in person at a regional office.

www.va.gov | 1-800-827-1000

Veterans Benefits Administration

For help filing disability claims, understanding your rating, or appealing a decision, the Veterans Benefits Administration provides guidance and representation through accredited Veterans Service Organizations (VSOs).

www.benefits.va.gov

National Veterans Legal Services Program (NVLSP)

A nonprofit organization that trains and supports advocates to help veterans and their families obtain the benefits they have earned. Provides free legal representation in VA appeals.

www.nvlsp.org

Veterans Crisis Line

If you or a veteran you know is in crisis, confidential support is available twenty-four hours a day, seven days a week.

Dial 988, then press 1 | Text 838255 | www.veteranscrisisline.net

National Center for PTSD

A leading source of research, education, and clinical resources on post-traumatic stress disorder in veterans and civilians. Includes self-help tools, treatment locators, and resources for families.

www.ptsd.va.gov

American Legion | Veterans of Foreign Wars (VFW)

Both organizations offer free claims assistance, legal advocacy, and peer support through local chapters nationwide. They are among the most established VSOs in the country.

www.legion.org | www.vfw.org

About the Author

Dr. Kara Lock-Harris is an author, speaker, and entrepreneur whose work sits at the intersection of faith, justice, and the power of story. As the founder of KLH Business Solutions and Letters From Christ, she serves leaders, visionaries, and everyday people who are ready to turn their lived experience into work that heals, informs, and endures.

With a background in business consulting, ministry, and community leadership, Dr. Lock-Harris brings a rare combination of scholarly rigor and spiritual depth to everything she writes. She specializes in ghostwriting, manuscript development, business plan writing, grant writing, and faith-based content creation—helping clients transform vision into language and language into legacy.

Battle Hymn of the Disregarded is her most personal work to date—not because it is her story, but because she was chosen to carry someone else's. She received Lewis Cain's message on a Sunday morning while sitting with her Bible and a cup of coffee, and she understood immediately that the assignment was not accidental. She accepted it.

Dr. Lock-Harris is a member of the American Business Women's Association and resides in the Milwaukee area.

To inquire about speaking engagements, ghostwriting services, or business consulting, visit www.klhbusinesssolutions.com.

www.ingramcontent.com/pod-product-compliance
Lightning Source LLC
LaVergne TN
LVHW020234210926
848481LV00016B/1954
9798993015606